Hilda Simon

The Date Palm
Bread of the Desert

Illustrations by the author
Photographs 1910-1914 by Henry Simon

DODD, MEAD & COMPANY · NEW YORK

1 2 3 4 5 6 7 8 9 10

Library of Congress Cataloging in Publication Data
Simon, Hilda.
   The date palm, bread of the desert.
   Includes index.
   SUMMARY: Describes the history, importance, and
biology of the date palm and notes the requirements for
its successful cultivation.
   1. Date palm.   [1. Date palm]   I. Simon, Henry,
1884-1957.   II.   Title.
SB364.S54   584'.5   77-14244
ISBN 0-396-07523-1

To my father
upon whose notes and diaries this book is largely based
and without whom it would never have been written

Acknowledgments

It is always a pleasure for an author to acknowledge the help received while working on a manuscript. Although much of this particular book's contents is based upon my father's notes and observations, a great deal had to be added to make the story of the date palm complete. Not only does this tree's history span some 5,000 years; it also ranges from archaeology to plant cultivation, and from geology to different national legends and customs. The list of persons who contributed to the wealth of information contained in this book is therefore varied, and I am grateful to all of them.

Overseas, my thanks go especially to my former teacher, the professor of history, Dr. Herbert Koch, of Wedel, Germany; to Dr. J.E. Reade, Research Assistant in the British Museum's Department of Western Asiatic Antiquities; and to Carol A.R. Andrews, Research Assistant in the same museum's Department of Egyptian Antiquities. In New York, I would like to thank Mrs. Herbert Shrifte, of the Vanguard Press, and the officials of the American Numismatic Society. I also greatly appreciate the efforts of Dr. Rosa M. Cabrera, professor of Spanish at the State University of New York in New Paltz, to provide me with information about date palm history in Spain, Cuba, and Mexico.

In California, I was aided by the late Roy W. Nixon of Indio, who for many years was associated with the U.S. Department of Agriculture, and by Lieutenant-Colonel James P. Westerfield of Mecca, California. The latter managed to arrange for me such fascinating interviews as that with Mr. Virgil Lawson, chief of the Desert Cahuilla Indians, and with the daughter of the early "date pioneer," W. L. Paul.

My very special thanks, however, go to Mr. Kenneth O. Shoots, a long-time resident of Indio, who proved to be an invaluable guide during my research excursions in the Coachella Valley. Mr. Shoots not only knew every road, lane, and byway in the valley, he also was ready to venture where there were no roads at all, even at the constant risk of getting stranded in the sand. I feel that the enthusiasm and interest he displayed for my undertaking, as well as the great patience, perseverance, and zeal with which he devoted himself to it, deserve an extra measure of appreciation.

—Hilda Simon

Contents

Illustrations

The Date Palm
Bread of the Desert

The road to Mecca, California, in 1910.

1. Pilgrimage to Mecca

Author's Introduction

E VER SINCE I can remember, I have felt a very special affinity for the desert, even though I knew it only from hearsay until quite recently, when I set eyes upon it for the first time. As a child, during the many years we lived in Europe, where the weather was mostly cool and wet, and even the summer season short, I spent countless hours daydreaming about the desert as I knew and pictured it from my father's descriptions and photographs. In those wonderful childhood dreams, in which everything was possible, I would mount my horse—nothing less than a magnificent Arab steed, of course—and ride through the hot white sand of a valley framed by multicolored hills outlined against a deep blue sky, and dotted by stately palm groves and settlements with such Middle Eastern names as Oasis, Arabia, and Mecca—especially Mecca. But "my" Mecca was not the fabled, forbidden Holy City of Islam, the goal of every *hadji*, or Moslem pilgrim. Instead, it was a small, little-known and—except in my dreams—rather undistinguished community in a southern California desert valley, where my father had settled years

Henry Simon displaying a live gopher snake prior to releasing it unharmed.

before he got married and started having a family. Not far from Mecca, on what was at that time still wild and untouched desert land, he lived alone miles from the nearest human habitation, and had for company only the coyotes that howled near his cabin at night, and the wildcats and rattlesnakes that hunted their prey in and around the mesquite thickets in the vicinity.

My father initially decided to spend time in the desert for reasons of health; several years earlier, a lung ailment had forced him to break off his studies at Oxford University, and seek a warm climate. Encouraged by some American friends he had met in England, he chose the southern California coast. Although his health improved considerably in the first years, a doctor friend who knew the desert believed that a prolonged stay in an arid region would be even more beneficial.

Though he thus had come to the desert for a very special reason, my father stayed on much longer than necessitated by his health because he fell in love at first sight with the unique—and at that time still un-spoiled—beauty of that particular part of the low-lying Colorado Desert, as this arid region of southeastern California is called. Despite the fact that life then was harsh and primitive, and lacking in all conveniences, he felt an exhilarating sense of freedom with so much uninhibited wilderness around him; all his life he dreaded closed-in spaces. To him, the "wasteland" never seemed monotonous, lonely, or desolate. The changing moods of the day reflected in the shifting colors of the distant hills and mountains, the brightness and clearness of the moonlit nights, the majestic stillness of the desert were sources of pleasure to him, a pleasure that was heightened by the great improvement in his health he soon experienced. Being a nature lover, he also found the wealth and variety of plant and animal life in this arid region ever fascinating, and spent long hours observing the wild creatures around him. He admired the intelligence and resourcefulness of the coyotes, and was perhaps the only settler in the valley who never killed—or tried to kill—one of these small wolves. He tamed horned toads and swifts to the point where they would climb up on his knee to take food from his fingers, and had "pet" kangaroo rats that visited him nightly and woke him up for a handout of rice. The notes and observations on animals in his reminiscences are a gifted naturalist's vivid and accurate picture of the wildlife as it existed in those regions more than half a century ago.

The doctors had advised him to be active out-of-doors as much as possible. Looking around for something to do, my father decided to

Assembling the well rig near Mecca in 1911.

Henry Simon checks the 40-foot rig.

apply his considerable engineering and mechanical skills to the fine art of well-drilling—water well-drilling, that is, for getting enough water was the prime concern of all settlers in the desert. Using special heavy oil-well equipment in place of the then common, much lighter apparatus, he soon had on his property a deep well capable of supplying large amounts of top-quality, nonalkaline water. This aroused the interest of the owners of the adjoining tracts of land, who at that time, in the years just before World War I, were getting ready to try date palm importation and cultivation on a large scale, a project that called for proportionately large amounts of irrigation water. My father quite naturally became interested and involved in these plans, and before long accepted the company's proposal to go to North Africa as their representative on a date palm-buying trip. After successfully completing that difficult and at times risky mission, he stayed in the Coachella Valley for another year to supervise the irrigation system he had designed for the newly established nursery, which was supplied with water from his well.

During his journeys in North Africa, and during the period of his involvement with date palms in California, my father made intensive studies of this tree, its history, biology, and the requirements for its successful cultivation. This research was facilitated by his lifelong interest in Middle Eastern history and culture, which had led him earlier, while still at Oxford, to take up the study of Arabic, the knowledge of which stood him in good stead during his travels in Africa.

Almost from the very beginning of his involvement in the date plantation project, my father envisioned writing some day about this tree and its history, both of which increasingly fascinated him. His plans are evident from his copious "Notes on the Date Palm" made during his African trip, as well as from the hundreds of photographs taken by him both during that journey, and later in the California desert. Many of the original prints of photographs in which he appears are still marked on the back, in his own handwriting, with captions such as "Author Henry Simon with the first load of [palm] offshoots," or "The author having coffee with an Arab date grove owner." The planned book never materialized, however; circumstances beyond his control, and especially the crippling Parkinson's disease that plagued him for the last seventeen years of his life, thwarted that and many other plans.

My early interest in the desert was rekindled when, in sorting through his many more or less finished manuscript drafts and notes after his death, I found the material on the date palm. I promised myself then that I would some day write the book he had planned. However, the time, funds, and freedom to travel and do the research necessary for that venture did not exist then, and it took quite a while for me to reach the point where I could afford them. A few years ago, however, I finally set out for the desert valley and the Mecca of my childhood dreams.

Enjoying a brief work break on the well.

My first view of the scenery that heretofore I had seen only in my father's old photographs came more than six decades after he had set eyes upon it for the first time, and indeed it had changed greatly, as he repeatedly had warned me it would. There are many highways and byways crisscrossing the valley now, and there is not much left of the desert as he knew and described it, for most of it has long since been turned into cultivated land. There is even less left of the way of life as it existed then. Modern conveniences have made life easy and comfortable for the vastly increased human population, with a proportionate decrease in the native fauna and flora. The coyotes, wildcats, rattlesnakes, and kangaroo rats have been dispossessed, and along with mesquite, sagebrush, and cactus, have retreated into the more remote areas of the surrounding hills and canyons. It is all so very much changed—and yet, as I gazed upon the valley, it took very little effort to bring back in my mind the scenes of yesteryear, for much of the essential beauty is still there. The background of the painted hills still frames the valley, and the majestic mountain peaks, capped with snow throughout much of the year, are outlined against the sky now as they were then. And if one leaves the roads and ventures a little into what is left of the desert, that also still is much as it was years ago. I saw flocks of California quail, the strange roadrunner scurrying along in search of prey, and a bright carpet of verbena and other wildflowers bloom among the sand dunes in the spring. I also saw the sight my father was not destined to see: large groves of towering date palms along the highways and roads every-where, an impressive legacy of the pioneering men of many decades ago.

In order to fill the gaps in my father's notes, and include the present state of date cultivation in the valley as well, I had to do a great amount of research, and seek information from many sources and individuals. The wealth of facts and facets about the tree's ancient history emerging from these studies led me far beyond the interest I had felt for the subject in the beginning. Not only was I able to understand why my father had been so intrigued by this palm; I also found myself viewing a large part of the history of Western civilization framed, as it were, by the feathery evergreen fronds of the date palm. This was a novel experience not only because I learned much about ancient cultures and their customs; it also opened up a different perspective—and one far more peaceful and pleasant than those I had heretofore known—of man's so frequently blood-drenched and violent progress through history.

Here, then, was a story, not of war and strife, of conquest and slaughter, of cruelty, callousness, and inhumanity to man and beast, but rather of planting and tending, preserving and propagating the palm that, with good reason, was known as the Tree of Life. Only occasionally in this tale does one get a marginal glimpse of the violence that has marked man's recorded history everywhere; for the most part, it is the story of an enterprise that has harmed no one, and helped many. The ancient laws and religious ceremonies pertaining to the date palm are benevolent. Priests and even gods, depicted on the artifacts of long-gone civilizations as playing a part in this story, are not engaged in bloody rites of human or animal sacrifice; instead, they are seen watering, tending, or pollinating the palms, or offering dates as an oblation.

As cultivation of the palms spread ever further through the ancient

View from the well site in 1912 . . .

world, it brought life to the inhospitable wastelands of the great Sahara Desert, providing food and wood, shade and shelter and fuel for the inhabitants. After thus becoming the very basis of life in vast desert regions, the ancient symbol of victory was eventually established across the oceans in the New World.

The story of the tree traced in these pages spans six thousand years and many more miles, linking ancient civilizations with present-day California, where the direct descendants of the palms that witnessed the splendor of Babylon and the pageantry of the Pharaohs are cultivated today very much as they were in Nineveh and Memphis thousands of years ago.

In a way, obtaining all the facts about the date palm's most recent history was more difficult than researching its past. Except for the captioned photographs, my father left few written records of the California date plantation, probably because he was so thoroughly familiar with every detail that he felt no need for notes. The tasks of closing the remaining gaps fell to me, and obtaining reliable information was difficult as well as time-consuming, because the people involved in that project were long since gone. Thus locating what had been my father's property, and the exact spot where he drilled the well that supplied that first nursery, took countless hours of circling a six-mile area and comparing mountain outlines with those on the old photographs, until every peak and dip fitted those on the pictures exactly, and I could verify that the old well was indeed there, covered up and in the

. . . and sixty-five years later, in 1977.

Palms imported from Algeria in 1913 bear fruit on a private ranch near Mecca in 1977.

middle of what is now a vineyard. Finally, however, all necessary information was assembled, and the book could be completed.

I realize, of course, that it cannot be the same as if my father had written it, for apart from everything else, there is a quality of immediacy that distinguishes all accounts of personal experience. I have tried, however, to keep as much as possible of that quality by quoting frequently from his travel journals.

About half the illustrations in this book are photographs taken by my father—or of him at his direction—in North Africa and California between 1910 and 1914; the captions are in many cases his own. The rest of the illustrations are for the most part either my own drawings, or photographs taken by me during my research trips to California. The pictures of ancient artifacts and architecture are courtesy of several museums. I trust that this combination of the old and new in both text and pictures will be successful in conveying some of the romance and adventure connected with the date palm's ancient history, as well as with the chain of events that helped establish the "Tree of Life" in the New World.

Assyrian soldiers in a vanquished enemy town's date grove in Babylonia.
(Sculpture from the palace of Sennacherib, Nineveh; seventh century B.C.*)*

2. The Dates of History

Food from the Garden of Eden

"THOU ART CREATED of the same material as this tree which henceforth shall nourish you," the Archangel Gabriel told Adam in the Garden of Eden. With these words, according to Moslem religious tradition, God's heavenly messenger and servant designated for all time the date palm as newly created man's chief source of food. The most widely told version of this ancient Islamic tale asserts that Adam, at God's bidding, cut his hair and his fingernails and buried the clippings in the ground. Immediately, there sprang up from that spot in the Garden a fully grown palm tree, heavy with clusters of ripe and delicious dates. Faced with that miracle, Adam sank to his knees and worshipped God, whereupon the Archangel Gabriel appeared and spoke the words quoted above.

Shortly thereafter, the story continues, Satan appeared and asked Adam why he was worshipping a tree. Upon being told what had happened, Satan became so infuriated at this sign of God's beneficence to man that he wept tears of rage. As they moistened the palm's roots, they caused the tree to bring forth the sharp, formidable spines that to this day grow at the base of the leaves.

The ancient tale was confirmed and augmented by the Prophet Mohammed, who consecrated the date, and admonished his followers to honor and acknowledge that incomparable tree as man's "maternal uncle," asserting that, among all trees, it is the "one blessed, as is the Faithful [Moslem] among men." The Prophet claimed that the close relationship between the palm and man is clearly proven by several characteristics shared by both. According to him, these include not only the fact that both have two sexes, but also that the palm perishes when exposed to too much strain, and dies promptly when its head is cut off.

Such tales, of which there are many, clearly indicate that the tall, stately date palm with its slender trunk and crown of graceful feathered leaves must have an ancient and romantic history. The facts support this assumption; the tree's record can be traced back to the dawn of Western civilization. It is one of the oldest cultivated trees in the world, and may well be *the* oldest, although such absolutes are not easy to establish.

To people from other parts of the world, it may be somewhat difficult to grasp the full significance of what the date palm has meant to the peoples of the Middle East and the Sahara Desert, who for excellent reasons valued this tree more highly than any other. In many regions, dates in the past took the place occupied by such staple food crops as rice, wheat, or potatoes in other cultures. The ancient association of the Semitic race with this palm is a unique record of large populations depending upon a single species of tree for much of their sustenance, and also immortalizing it in their religious, literary, and artistic traditions over a span of many centuries. This record has no equal anywhere else in the history of mankind, for even the versatile coconut palm, vital as it was—and to some extent still is—to the economy and culture of the Polynesians, cannot compare to the overwhelming importance of the date palm to the Semitic populations of the Middle East and North Africa. Like a thread of peaceful, life-giving green, the concern with this tree and its cultivation runs through the almost six thousand years of a history that was as distinguished and fascinating as it was turbulent and beset by wars and violence.

Despite intensive research into the origin, biology, and habitat of

Date groves and irrigation basins in the city of Madaktu in Elam. (Sculpture from the palace of Sennacherib)

this tree, no one today can pinpoint with certainty its original home, nor do any "wild" date palms exist. The fossil record shows it to have been widely distributed throughout the entire Mediterranean region during the Eocene period, but there is no historical record of wild date palms similar to that of many other important fruit trees. The earliest recorded information—and it is very early in man's history indeed—already describes the palm as a cultivated tree only. Dating from civilizations that flourished as long as six thousand years ago, these records tell of date cultivation in ancient Mesopotamia—the region which today lies largely within the borders of modern Iraq—but give no indication at all of where those trees originally came from. Nor do they tell us how, and

by whom, the idea of cultivating them, which involves a rather compli-
cated process, was first conceived.

It seems likely, however, that the home of the date palm, at the time
when man appeared on the world scene and began to make history, was
in fact the lower Euphrates-Tigris valley, the region so rich in records of
civilizations that were already ancient when the first pyramids were
being built that it is often referred to as the "cradle of Western
civilization." In Semitic religious legend, that fertile valley was iden-
tified as the location of the Garden of Eden, where mankind's history
began. Some experts lean to the theory that the date palms cultivated
there originally came from Arabia; fossilized remains of date palms
from the Eocene period do not clarify that particular point but attest to
the tree's very ancient ancestry. In that respect at least, they do not
conflict with Moslem religious tradition, which regards the date palm as
the legendary "tree of life" mentioned in the story of the Genesis.

Botanists believe that date palms probably grew wild in prehistoric
times along the ravines bordering the desert regions of Mesopotamia,
where the climate was sufficiently hot and dry to secure ripening of the
moisture-sensitive fruit, but where the ample amounts of water the roots
of this tree need to absorb at least periodically also were available.

Clay tablets from the ancient Assyrian and Babylonian civilizations
describe the culture of the date palm in some detail. A number of
Assyrian monuments, bas-relief sculptures, and other works of art
depict stages of cultivation, and especially the pollination, by man, of
these trees. Some sculptures not only show entire date groves with
fruit-laden palms, but also the trenches and basins by means of which
the necessary water was conducted into the groves.

So highly valued was the tree by these ancients that it was regarded
not just as an economic asset, but as a sacred plant used in religious rites
and ceremonies. In one bas-relief scene, an Assyrian god is shown with a
pail in one hand and fertilizing a date palm with the other.

Sumer, the region in the lower Euphrates valley inhabited by the
non-Semitic Sumerians for centuries before the Babylonians, appears to
have been the center of ancient date cultivation. In that respect at least,

times have changed but little over the past five thousand years, for the Bassorah-Mohammerah district on the Shatt-el-Arab—the Euphrates-Tigris delta—today still has some five million date palms. In the ancient days, the deciding factor in choosing that particular region for date cultivation was probably the ease with which the groves could be irrigated, for high tides raise the water level as much as six feet above normal, thereby causing a flooding of the groves without any human intervention or labor.

The most detailed, vivid, and interesting ancient records on the date palm and its cultivation can be found in the famous Code of Hammurabi, that astonishing and well-organized collection of laws and edicts published by Hammurabi, the greatest king of the first Babylonian dynasty. This extensive work of legislation, which dates from approximately 1950 B.C., is based upon much older Sumerian legal material. The first fragments of the stone pillars inscribed with the cuneiform characters widely used in that period were almost four thousand years old when they were discovered right after the turn of the century in 1901-02.

The portions of the Code pertaining to the date palm—the laws regulating the sale and renting, as well as the cultivation of the date orchards—are extensive and detailed. We learn, for example, exactly how a landowner who preferred not to cultivate his own orchard, and a man who had no land of his own, could both arrange to get equal shares in annual date harvests without the outlay or investment of great sums of money. In this arrangement, the landlord would simply lease his land to a tenant farmer, and give him access to young palm suckers, or offshoots, suitable for planting. Although the owner then had to wait a full four years without getting any return on his land, he did not have to concern himself with its cultivation, for during this period, the farmer—or gardener, as he was called, for date groves were considered gardens—was solely responsible for the planting and tending of the trees. After the fourth year, with the palms presumably producing their first good crop, both landlord and tenant began to share equally in the harvests, the only stipulation being that the owner had first choice.

One-half of the date crop—even though it averaged about 100

Assyrian soldiers cutting down an enemy's palms near the city of Dilbat.
(Drawing of a lost sculpture from the palace of Sennacherib)

pounds per tree—may appear a rather meagre recompense for four long years of hard work. It would be hasty to conclude, however, that the tenant gardener was forced to work for nothing during the first years, for date orchards in the Middle East were then—and still are in many places—designed to yield a variety of secondary crops planted among the palms. Some of these crops grow better in the part-shade of the groves, where they are protected against the very hot sun, than out in the open. In the oases of modern Algeria, secondary crops may range from oranges, apricots, and nuts, to a variety of vegetables, including excellent asparagus; in ancient Babylonia, they probably consisted of plants such as sesame, grain, and clover, all of which the tenant had the right to harvest and keep for his own consumption, while waiting for the palms to come into bearing. The period of waiting—four years—mentioned in the Code clearly indicates that the young trees were not raised from seeds but were offshoots cut from the base of older palms, for seedling palms need from between eight to ten years to produce their first good crop.

The great value of the trees to those ancient civilizations is nowhere underlined more impressively than in the clause of the Code dealing with the punishment for any damage done to the date orchards. These punishments are neither brutal nor inhuman, and consist of money fines only, but the fines are so stiff that they would have been a strong deterrent. Thus cutting down a single palm without consent of the owner merited a fine of one-half *mina* of silver, a *mina* being approximately fifteen ounces. Half a pound of silver seems a rather heavy fine for the destruction of one tree even today. For the average citizen of that era, it must have been a prohibitively high sum, which of course is exactly what it was meant to be.

Even though date palms in Babylonia had to be planted about thirty feet apart, which gave each tree sufficient space but limited the number to some fifty per acre, date orchards were more than twice as valuable as any other type of cultivated land, regardless of what kind of fruit trees or other crops were planted on it. In wartime, this heavy reliance upon the palms could—and did—turn into the Achilles heel of any defeated people

Distribution of the date palm in the pre-Christian Middle East.

of that region, for destruction of their date groves was a fatal blow. Ancient Assyrian monuments and sculptures show numerous scenes in which victorious soldiers are cutting down a vanquished enemy's date palms. This was an effective way of destroying one main source of sustenance, causing serious food shortages, and thereby keeping the conquered poor and less likely to wage wars of retribution.

From Mesopotamia, the art of date cultivation quickly spread to the adjoining arid regions. In ancient Palestine, there were many date groves, even though conditions for good fruit production were not ideal along the coast. Today, date palms in those parts are found only around the lake of Genesaret, the Dead Sea, and Jericho, which in Biblical times had such large groves that it was known as the "City of the Palms."

References to the date palm abound in the Bible. We learn from the Book of Psalms that the "righteous shall flourish like a palm tree," and in the Song of Songs a beautiful woman is likened to a palm. According to the Book of Judges, the Jewish prophetess Deborah dwelt under a date palm, and later, while sitting under that "palm tree of Deborah," she judged the Tribes of Israel. The Book of Kings tells us that the date palm was one of the ornamental motifs used on Solomon's great temple, and a number of ancient Judean coins feature that tree.

The religious, symbolical significance of the palm in ancient Palestine was considerable and varied. Palm leaves were used in temple services during the Feast of Booths, and carried before kings as symbols of victory, because palm leaves are evergreen and do not fall off or wither during the winter season. They were featured in the great victory processions for King David and for Simon, one of the Maccabees, after he had captured Jerusalem. The hailing of Jesus with "palm branches" during His triumphant entry into Jerusalem a few days before His cruxifixion signified that he was greeted as a King come to conquer. The Christian holiday known as Palm Sunday, one week before Easter, is based upon that incident. On Palm Sunday, the priests in some Christian denominations still consecrate palm leaves, even though the ancient Jewish symbolical significance of these leaves has long since been obscured.

Two bronze coins depicting the date palm.
(From the 2nd Jewish Revolt, A.D. 132-135)

Two glass kohl (eyelid-coloring) tubes in the form of palm columns. (Late Eighteenth Dynasty; approximately 1400 B.C. and later)

The wealth of ancient literature about the date palm includes everything from agricultural data to poetry; Greek travelers returning from Babylon and Tyre described the vast date groves and their cultivation. A poet of ancient Persia waxes lyrical about the palm's beauty and usefulness, and proceeds to prove his point by listing 365 ways—one for each day of the year—in which this incomparable tree benefits man.

Because many of the timber-poor arid regions considered the palm's wood and fibers welcome building materials, its value was not limited to the fruit, even though that was by far its most important product. Dates were used in a variety of ways in antiquity, a favorite being a kind of "honey," or syrup, made from soft pressed dates. Biblical references to honey, such as the one about the "land where milk and honey flows," are believed by historians to refer to this "date honey" rather than to the product of honey bees.

Some scholars believe that the date palm was introduced into Egypt perhaps as early as five thousand years ago. There are excellent pictorial records from the Nile region that include Egyptian bas-relief sculptures and paintings—some over four thousand years old—in which stages of date cultivation are illustrated. A carving from the kingdom of Memphis pictures a priest irrigating palms, whereas a scene from the Eighteenth Dynasty (1580-1335 B.C.) shows King Thutmosis I offering dates as a sacrifice. The magnificent palm pillars of Fifth Dynasty architecture are famous examples of classic Egyptian art.

Egyptian monolithic granite column with a palm capital. (From a temple built by Rameses II, Nineteenth Dynasty; fourteenth century B.C.)

In Egypt, as in Assyria and Palestine, the palm was used in the religious symbolism of the country. Some stages of date cultivation—especially the pollination of the trees—were part of important sacred fertility rites.

The great success of date cultivation in ancient Egypt depended largely, as it had in Mesopotamia, upon the availability of large amounts of irrigation water. We know, of course, that a very efficient irrigation system was one of the great accomplishments of the ancient Egyptians, who had perfected an ingenious way of regulating the periodical floodings of the Nile, and harnessing this water for their own use.

In Egypt, as in the Persian Gulf region, the date palm has survived all the wars, revolutions, and other upheavals that beset this part of the Mediterranean region over the centuries. Today, the date groves of Egypt number between eight and ten million trees, and dates are still important to the nation's economy, as they are to other North African countries.

A hobbled caravan camel and its young during a rest period.

Camel caravan carrying palm offshoots across the desert.

The palm first became established in parts of the Sahara Desert nearly two thousand years ago, as Semitic tribes spread out and began to take over this vast region. Use of the camel as a beast of burden capable of transversing great stretches of waterless wasteland while carrying heavy loads and subsisting on meagre fare and a minimum of water enabled these tribes to settle in the desert oases. Only camels could transport the precious palm offshoots, weighing between fifteen and forty pounds each, which were needed to establish date groves in the desert. It had been known since ancient times that the most satisfactory way of propagating these palms—and the *only* way to reproduce a desirable variety—was to grow them, not from seeds, but rather by cutting off and planting the suckers, or offshoots, that form at the base of the parent tree.

Over a period of several centuries, date palm cultivation spread throughout the Sahara Desert, and was established in any oasis capable of providing the water necessary to raise these trees. Despite the constant battle against shifting sand dunes, dates soon became the most

A young date garden in the Sahara threatened by shifting sands.

important staple food item, the legendary "bread of the desert," for large populations in the arid regions of North Africa.

The myths and folklore of any people are rich sources of information about past customs, habits, and lifestyles; in the Middle East and North Africa, tales and legends involving the date palm are plentiful. The ancient veneration of the date was carried forward in a different form by the Islamic religion. As behooves a tree with so illustrious an ancestry, the palm and its fruit are credited with a number of outstanding and miraculous properties in Arab religious lore. The palm is even featured in one Arab tale about the birth of Jesus—whom Moslems respect as a legitimate if relatively minor religious figure.

In the story about Christ's birth, Mary was overcome by sharp labor pains; unable to take another step, she sank to the ground near the base of a palm tree. The pain became so unbearable that she prayed for death as a release. But at that moment, the child Jesus spoke to her from inside her body, bidding her to be of good cheer, and ordering her to shake the palm tree and eat the dates that would fall into her lap. Obeying the command of her unborn baby, Mary did as told. And lo! and behold! ripe sweet dates fell into her lap although it was not the season for them, and when she had eaten them the pain immediately left her.

Taking special note of this story, Mohammed advises all expectant mothers to nourish themselves with dates. He claims that such a diet is effective for increasing both the quantity and quality of the young mother's milk, and therefore is of great benefit to the newborn child, for whom the prophet considered that milk of vital importance. In his evaluation of the importance of mother's milk, Mohammed was far ahead of his time, for the most modern medical views tend to concur with him.

As can be expected from such a miracle food, the medicinal values of the date were allegedly many and varied. If one wants to believe old Arab medical lore, any ailment from fever to impotence can be successfully treated with a diet of dates. Mohammed himself made perhaps the most sweeping statement of all, declaring that anyone who eats seven dates of a specified variety early in the morning will be protected all day long against both "poison and treachery."

Nineteenth-century drawing of an oasis in the Sahara. (From the journals of German explorer Gerhard Rohlfs)

Belief in the special medicinal properties of dates was widespread throughout antiquity. In the first century A.D., the Roman author and naturalist Pliny the Elder wrote about the usefulness of the date as a tonic for a variety of ills. According to him, the fruit not only gives strength to sick persons weakened by long illnesses, but also is especially efficacious in treating chest pain and cough. The nugget of truth in his assertion is probably the fact that dates, which are rich in glucose and fructose, are a good source of quick energy. Other "authorities" recommend the sap of the palm for nervousness, kidney trouble, and as a tonic for the stomach. The male flowers, eaten as a salad with lemon juice, are supposedly a powerful aphrodisiac.

Since no rule can be without exceptions, there can be found in that general chorus of praise for the date a few dissenting voices, which darkly warn of the dangers of a date diet. However, in view of the fact that dates have been a staple food for millions throughout centuries, and that the most reliable sources have confirmed reports of people not only

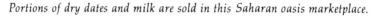

Portions of dry dates and milk are sold in this Saharan oasis marketplace.

subsisting, but remaining in excellent health, for more than six months at a time on a diet of nothing but dates fortified with milk, the few prophets of gloom and doom have not been very convincing.

Shorn of all fantastic claims, the truth is that dates are a truly valuable food. This applies especially to the drier, less sweet and sugary varieties known as bread dates to the Arabs, and traditionally preferred by them for much of their everyday use. Although dates contain only about 2 percent protein and the same amount of fat, they have appreciable quantities of iron, magnesium, potassium, phosphorus, calcium, and copper, plus small amounts of vitamins A, B1, D, and G. The invert sugars that make up most dates' entire sugar content are not acid-forming like cane sugar, and there is enough fibre, even though it is fine and soft, to provide sufficient roughage. The fact that people whose staple diet consisted of dates were not plagued by many common digestive problems was noted long ago in ancient Greek and Roman reports about the Middle East, and more recently by Western travelers in those regions.

Eaten with protein-rich foods—the habits of splitting a date, removing the stone, and filling the cavity with a dab of butter or thick sour cream is centuries-old in the Middle East—dates provide a very satisfactory nourishment. Although some of the traditional recipes featuring dates may not be attractive to the Western palate, many others are. Cottage cheese and dates, for example, make a very pleasing and satisfying meal, which has the distinct advantage of being nonfattening and yet supplies protein, vitamins, minerals, and invert sugar for quick energy, as well as the necessary roughage. Doctors specializing in modern dietary medicine are investigating the claim that a diet of this type, featuring a high percentage of dates, tends to help in lowering blood pressure.

The Prophet Mohammed, by the way, did not just give good advice to others regarding a date diet; he also preferred it for himself. During his lean years in Medina, when he was very poor and unable to buy other food, he often subsisted for weeks on nothing but dates washed down with water. Later, when he could afford more expensive foods and

From childhood onward, much of the "date Arab's" life is spent in his grove.

developed a gourmet palate, he still liked to eat dates in a variety of what were considered to be delicious combinations. Many of these—such as a salad made of dates and cucumbers, which latter allegedly "cool the fire" of the dates—may appear somewhat strange to Western tastes, although many Western travelers in those regions adapt themselves quickly to local food preferences. Mohammed's favorite date dessert, a concoction of dates, honey, and butter known as *khabis,* for which the Prophet called down Allah's blessings on the originator, is rich and very sweet, but according to all reports a thoroughly delectable tidbit.

Although dates must be supplemented by protein-rich foods such as milk, cheese, or meat for a balanced diet, the date palm for centuries was the very foundation of non-nomadic Arab life in the desert, a fact confirmed by the adventurous early explorers of the Sahara. The few intrepid Europeans—especially the two Germans Gustav Nachtigal and Gerhard Rohlfs—who crisscrossed that fearsome desert in the nineteenth century, brought back not only geographic information that eliminated some of the "white spots" marking uncharted areas on the contemporary maps of those regions, but also reliable reports about the desert's inhabitants and their way of life. Time and again only a hair's-breadth away from death, Rohlfs later lectured about his trips in many countries, including the United States, where he held audiences spellbound with the accounts of his daring exploits. One of his narrow escapes involved a lynch-minded mob of fanatical Moslems, who blamed the loss of three hundred precious date palms felled by a sandstorm on the presence of the "infidel," the "Christian dog," in their midst.

The vital importance of the palm to the inhabitants of the Saharan oases can be found in the fact that not just the fruit but every single part of the tree was utilized. In his "Notes on the Date Palm," which are based upon his observations of life in the North African oases in 1913, Henry Simon sums up the significance of the tree even in recent times. "The palm supplies food for man and beast; fibre for ropes and mats;

Arab child leads burro laden with dried date fruit stalks, which will be used for fuel.

Market day in a North African oasis.

leaf material for roofs, baskets, and hats; wood for building, irrigation dam work, and fuel; sticks for tents and kindling; and *lagmi* (palm wine) from the sap when the palm is dying. No part of the tree or its fruit is wasted; the better dates are used or sold as food, the more inferior fruit is fed to horses and mules, and even the stones are ground up and mixed with other ingredients for camel feed."

Among the great advantages of dates as food in hot climates is their resistance to spoilage. In the past, the only food an Arab could take along conveniently on many a desert trip was a bag of dry dates. A handful several times a day had to suffice for both the rider and his mount, although the former supplemented this diet with some meat or milk bought from passing nomads if and when available. The proven capacity of Arabian horses to endure long hours of carrying the rider and his belongings through the desert, and then being satisfied, if necessary, with nothing more than a few handfuls of dates and a drink of water at the end of the day, aroused the admiration of early European travelers. The German explorer Nachtigal noted in one of his books that without the camel, the date palm, and the Arabian horse, Arab life and culture in the desert would not have been possible.

Even though date palms and their fruit were used predominantly for the necessities of life in the Middle East and Africa, they were sometimes also converted into luxuries, including a few of questionable value. Thus in Baghdad, dates were used to make an arrack, a drink that reputedly had all the knockout power and aftereffects of the notorious French absinthe. Palm wine, on the other hand, is relatively mild; it is made from the sap that tastes somewhat like coconut milk in the

Date palm offshoots for sale in the marketplace.

unfermented state, and is not all that much of a luxury because the palms tapped for the sap are most often old and dying trees. The same is not true of the delicacy known as "heart of palm"—the terminal buds—which may be eaten raw, boiled, or else chopped and mixed with other ingredients as a salad. These buds are indeed a luxury because cutting them off means killing the trees.

Because of the prime importance of the palm to all aspects of Arab desert life in the past, the wealth of an individual was based largely upon the number as well as the variety of dates he raised in his gardens; this, of course, applied only to those Arabs living in the oases. There used to be a considerable difference, much of which has disappeared in the past fifty years, between the "date Arabs" and their way of life, and the Bedouins, or "Arabs of the mountains," who lived in tents, were constantly on the move, and counted their wealth in domestic animals such as camels, horses, and goats. Although many of these nomadic people also ate dates in quantities, they did not cultivate their own palms, but bought or traded the fruit they needed for meat, milk, skins, and other items in the oases through which they passed during their travels.

Because nothing is known about the origin of date cultivation, we also have no idea of how, when, and by whom the requirements for a good and ample fruit were discovered. For a variety of reasons, most of them related to the palm's peculiar lifestyle, this is not an easy or effortless undertaking, and demands know-how, constant attention, and at times much hard manual labor. Date cultivation cannot be compared at all with the growing of such fruit trees as pears, plums, and apples. Once the strain has been improved and established through selective breeding, these trees can be planted and then left to develop their fruit more or less on their own, with the necessary water supplied by rain, and pollination by insects such as bees. Many apple growers, for instance, insure failproof pollination by the simple expedient of renting beehives and placing them in their orchards during the spring. Even the harvesting of such trees presents no great difficulties, since most do not grow to excessive heights, and often can be kept low and bushlike through selective trimming.

Cultivation of date palms is a different matter entirely, and cannot be achieved by such simple methods, for here really nothing can be left to nature—except the ripening of the fruit by the sun—if an ample, top-quality crop is desired. Man has to attend to everything, from special care while planting the young, rootless offshoots and protecting them against the hot sun and wind, to irrigating and pollinating the mature trees, and even sometimes shielding the fruit against an occasional rainfall during the long ripening period. Harvesting of the crop also can be quite a problem with trees that cannot ever be trimmed to prevent them from growing excessively tall, and that bear their fruit in the crown at the top of an often forty- or fifty-foot trunk. But once established and properly treated, date palms will bear fruit—several hundred pounds of it—year after year for many decades.

Those who know the date palm well, like the Arabs, sometimes claim the tree has a definite "personality." Be this as it may, few would deny that this palm is well worth knowing, as much for its ancient and illustrious history as for its interesting biology and lifestyle.

Arab date grove owner at the entrance of his garden.

3. The Tree of Life

Growth and Lifestyle of the Date Palm

Among the words used to describe the palm which throughout the centuries has been one of the mainstays of entire desert civilizations there is a recurrence of adjectives such as "regal" and "stately." Gracefully proportioned, the tree's slender trunk fans out into a symmetrical crown of long, feather-like leaves at the top. Although most varieties do not grow much over sixty feet tall, date palms may attain a maximum height of one hundred feet, and an age of up to two hundred years. The leaves, which range from ten to twenty feet in length, have a grayish- or bluish-green color, depending upon the variety; their stalks are long and armed with sharp, hard spines—actually modified leaflets—at the base.

All palms, including the date, belong to a group of plants known as monocotyledons, which include the grasses, lilies, and orchids. The common feature of this otherwise highly diversified group is the single original leaf developed by the seedlings. These first leaves hardly ever resemble those of the mature plant. The date palm seedling, for instance, looks much more like a robust grass during the first year or so of its life than like a future palm.

The monocotyledons count among their ranks some very unusual, interesting, and "individualistic" members. The orchids with their multitude of fantastically varied and complex forms and lifestyles are a case in point.

Seedling date and Washington palms.

Although they cannot compete with the orchid family either in numbers of species nor in diversity of appearance, the Arecaceae, as botanists have labeled the palms, can boast highly individualistic members with a wide range of shapes, habits, and habitats. Palms may vary in size from the tiny *Mallortiea,* which do not exceed a height of two feet, to the two hundred-foot towering giants of the genus *Ceroxylon* native to the Colombian Andes. The trunk of some mature palms may measure only a few inches in diameter; others attain a thickness of several feet. Certain palms are vinelike creeping plants that may grow to a length of five hundred feet, and still others, such as the well-known saw and etonia palmettos of the American Southeast, are bushy or shrubby. Similarly, they may occur in a great many different habitats. Although the great majority are at home in low-lying regions, some are found in mountains, and a few grow at altitudes up to 14,000 feet above sea level. Certain species prefer humid localities with heavy rainfalls; others like warm but dry habitats, and still others thrive in the sandy soil of the seashore. The fruits may range from a berry the size of a pea to the huge, 40-pound sea, or double, coconut, the largest single fruit produced in the entire plant kingdom.

Despite these astonishing variations, all members of this group share certain characteristics setting them apart from other trees. For one thing, palms typically have no branches, only very large and usually hard, stiff, and tough-textured fan- or feather-shaped leaves. However, branching does occur in a few, notably in the group that includes the so-called gingerbread tree, the African doum palm. The popular name of this large fan palm stems from the pulp of the fruit, which has a flavor resembling that of gingerbread. These doum palms often have naturally branched, or rather forked, trunks; in that, they are somewhat like the peculiar "Joshua trees"—actually giant yuccas—of the American Southwest which, like palms, belong to the monocotyledon group. In most other palms, branching occurs only as the result of an injury.

The exterior of the typical palm's slender, unbranched trunk may be smooth, rough, or even spiny. It supports a handsomely proportioned and symmetrical crown of large evergreen leaves. As new leaves grow,

The "Joshua tree," a giant yucca, is related to the palm group.

the older ones die and often drop off. In some species, however, the dead leaves hang on almost indefinitely and form a kind of shag around the trunk that protects the growing point and changes the appearance of the palm.

The stem structure of a palm differs considerably from that of the familiar timber tree. There are no annual growth rings, from which the age of the average tree may be estimated. A palm's growth in diameter does not continue year after year by the addition of new layers of cells; actually, the girth characteristics of a given species are determined in the early stages of growth. Palms have only a single main growing point; if that bud—it's "head"—is cut off, the palm dies.

The heavy-walled cells that conduct water and food are scattered among thin-walled cells resembling pith. The number, density, and arrangement of the heavy-walled cells determine the properties of the palm's wood. If they are concentrated toward the outer rim, the palm trunk has a ring of very hard wood surrounding an often soft, pithy interior, which in some species is valued almost as much as the fruit because of the starch stored in the trunk or the sugar content of its sap.

All palm leaves have one of two distinctive shapes, one of which resembles a fan, and the other a feather. Botanists call the former palmate, and the latter pinnate. The longest palm leaves measure

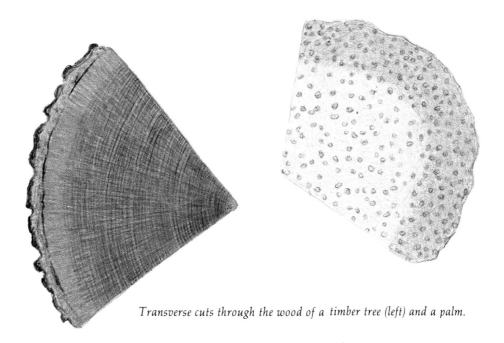

Transverse cuts through the wood of a timber tree (left) and a palm.

Leaves of a fan, or palmate, palm, and of a feather, or pinnate, palm.

sixty-five feet from base to tip; the broadest fan-shaped leaf is about sixteen feet across. In contrast to these huge leaves, the flowers are generally quite small, and often look like miniature short-stemmed lilies. They frequently make up in number what they lack in size, occurring in huge, dense clusters numbering hundreds or even thousands of individual blossoms. The fruit, which may range from a small, soft berry to the large, heavy, hard-walled sphere of the coconut, almost always contains only a single seed.

Of the approximately 4,000 different species of palms, about 1,200 grow in the Western Hemisphere, with Brazil and Colombia harboring not only the greatest number, but also the giants of the entire clan, the so-called wax palms of the Andes.

Only fourteen species of palms are native to the United States. Many of these are palmettos, which literally means "little palm," although some palmettos may grow to a height of eighty feet. One of the best-known is the cabbage palmetto, the state tree of South Carolina and Florida, whose edible bud is canned and sold as "heart of palm." Even more useful are the palm's fibres, which are made into brushes and whisk brooms that are unaffected by hot water and caustic ingredients.

Only a single native North American palm occurs in the West. This is the famous California *Washingtonia,* popularly known also as California or Washington palm. The scientific name of this species, chosen in honor of our first President, is *Washingtonia filifera.* Together with the closely related *Washingtonia robusta* of Mexico, it makes up the entire native palm population of southeastern California, Arizona, and Baja California. In those parts, the distinctive palms with their fan-shaped leaves are a common sight, usually as ornamental trees along streets and avenues or in gardens, but also growing wild in some arid regions. The wild groves, especially in the Coachella Valley of California, all have distinctive names—Hidden Palms, Twenty-nine Palms, Twelve Apostles—and are favorite tourist attractions.

The fruit of the Washington palm is a small, half-inch black berry with sweet, edible pulp and a single seed. These fruits were used—both in the fresh and dried state—as food by the Indians of the arid regions where the palms occur. In the ability to benefit man by what they contain or produce, the Washington palm is typical of its family. Wherever palms grow in the world, they always include some species valued for economic importance. One kind may have been sought after for its wood, another for its sap, and still others for the soft, edible pith, or for the oil they yield. Most important in mankind's history, however, have been the two species cherished for their fruit: the coconut palm in Asia, and the date palm in the Middle East and Africa.

Wild-growing Washington fan palm.

Members of the small genus *Phoenix*, date palms number only about a dozen species. Of those, only *P. dactylifera*, the common date palm, and the Indian *P. sylvatica*, which is cultivated for its sap, are of economic importance. The Canary date palm, *P. canariensis*, a close relative of the common date palm, is a graceful tree often used for ornamental plantings, but bears small, inedible, yellowish fruit.

The specific name of the common date palm, which means "finger-bearing" in Latin and from which the word date is derived, seems to refer to the oblong shape of the fruit. Some scholars, however, incline to view the Latin designation as an alliterative term, pointing to the similar-sounding Arabic *daqlan*, which means "seedling," as being too much of a coincidence.

No one seems to know very much about the origin of this group's generic name except that *phoinix* was the ancient Greek word for date palm. However, it also was the word for the phoenix bird of myth and fable, which, according to ancient Egyptian religious beliefs, rises rejuvenated from the fire into which it voluntarily plunges itself. The word *phoinix* actually means crimson, from *phoinos*, blood-red. What the connection is—if indeed there is any—between the color, the bird, and the date palm is not clear. Some scholars believe that the Greek associated the date palm with the ancient country of Phoenicia, and that the name for the tree arose in that way, but others maintain that it could have been the other way around, and that the country was named for the palms that grew there in numbers, which leaves us right back where we started from. There is even one group of experts who think that the palm's Greek name may have been inspired by the bright reddish or purplish color of certain varieties of dates during one of the ripening stages.

Date palms are dioecious; that is, the trees have separate sexes, and are either male or female. For ornamental plantings, the sex does not matter, for they differ very little in general growth characteristics and appearance. When grown from seeds, the sexes are about evenly divided. Because a single male tree yields sufficient amounts of pollen to fertilize an entire "harem" of female trees, it is uneconomical and

Crown of a young female palm in full bloom.

undesirable for a grower to have large numbers of male trees. In most instances, a ratio of approximately one male for every forty or fifty female palms is maintained.

Date palm flowers of both sexes are very small compared to the size of the tree; like those of most other palms, however, they make up in numbers what they lack in size. The waxy, cream-colored, tightly bunched male flowers are somewhat larger, more attractive, and more numerous than their whitish, globular, widely spaced female counterparts. The flowers are borne on long, slender strands, and are sheathed inside a woody spathe which opens to release clusters bearing hundreds of blossoms.

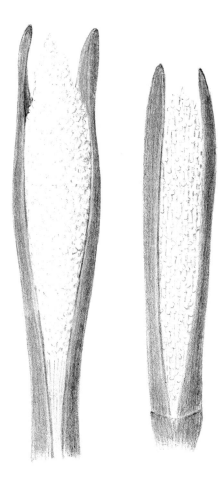

Male and female date flowers in their half-opened spathes.

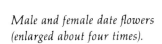

Male and female date flowers (enlarged about four times).

Two offshoots growing from the base of a parent palm.

In their natural state, date palms were undoubtedly fertilized by the wind, which carried enough of the dust-fine pollen from male to female trees to insure development of fruit in numbers sufficient to propagate the species. This kind of random fruit production was no drawback because the date palm, unlike its relatives, has another and most effective way of propagating its kind in the offshoots, or suckers, that normally grow on the parent tree. Possible failure of one type of reproduction among wild trees was offset by this "insurance." If left unattended, a proliferation of offshoots growing from the base of the parent tree will in time transform that palm into a huge, bushlike mass, in which first-generation offshoots may in turn grow their own suckers.

The palm's haphazard fruit production, which must depend largely upon the wind, is quite satisfactory from the viewpoint of continued survival of the species, but highly unsatisfactory to humans desiring a maximum harvest of the valuable and nourishing fruit. At some unknown but very early point in history, man discovered that pollinating the female trees by hand greatly increased the annual fruit yield. Even the most ancient records show this mode of pollination as an established system, which of course means that the significance of the difference between male and female trees, and the fact that pollen introduced into the female flowers results in fruition, must have been known to people living some six thousand years ago.

The success of ancient date groves also proves that these people—probably the non-Semitic Sumerians who preceded the Assyrians in Mesopotamia—evidently had grasped the fact that propagation of the palm trees could be effected better and more economically by cutting off and planting the offshoots, than by raising palms from seeds. Although the first full harvest may be advanced as much as five years by offshoot planting, this system involves a good deal of knowledge about the palm, for offshoots are produced only under certain conditions. There is a definite correlation between fruit production and offshoot growth; any increase in one demands a proportionate decrease in the other. To get good offshoots from a palm, many of the fruit clusters have to be removed, for the tree does not have enough strength for both a large fruit yield and the growth of offshoots. In any case, a harvest will consist of better quality fruit if excess clusters—some palms may have as many as twenty-five—are removed to give the others a better chance to develop. If at least one offshoot is left on the tree, it will continue to grow others, but if all the suckers are cut off, the palm will never again produce an offshoot.

Worker in a North African grove with male date flower sprigs.

From the time of ancient Assyria to our day, the method employed in pollinating date palms has remained essentially the same: as the flowers open, the strands of male blossoms are cut off, and men carrying them climb up into the crown of each female tree—no easy undertaking if the palm is tall—and tie a few of the strands inverted into the female flower clusters. The most modern way of pollinating the trees as practiced by many of today's growers in California actually differs very little from the ancient ways, except that the pollen is first applied to wads of cotton, which are then placed between the strands of the female clusters. At pollination time, lightweight aluminum extension ladders permit the workers to reach the crowns of even the tall palms without too much difficulty. In an effort to facilitate pollination, some growers have experimented with insecticide dusting apparatus as a means of applying the pollen. Regardless of the method, however, it still has to be done by man for each tree.

During experimentation with date cultivation in California around the turn of the century, plant biologists of the American Agriculture Department made a surprising discovery: good date crops could be achieved through pollination of the female flowers with the pollen of other palms. Thus excellent results were noted in date palms fertilized with pollen from the inedible Canary date. Even more surprising was the successful impregnation of female date flowers with pollen of the California Washington palm, which belongs to a different genus. Such experiments led some botanists to believe that the male date palms are unimportant for cultivation, and that the female palms alone will decide the quality of the crop. In the course of continuing experimentation, however, it became clear that the quality of the male palm does in fact matter over the long run, and that healthy, vigorous male date trees are important for maintaining stable fruit quality.

Dates need about six months to ripen, which seems long for such a small fruit. In the Old World, four distinct and different stages of ripening are recognized, and each has its own designation. Arabs distinguish *kimri, khalal, rutab,* and *tamar* in describing these stages. The dates are green in the *kimri* stage, but turn a yellowish or reddish color in

A strand of dates in four ripening stages.

the *khalal* stage, in which they also attain their maximum size. In the *rutab* stage, the date loses its bright coloring, and begins to ripen and soften. Finally, when it has dried out somewhat and is fully matured, it is *tamar,* which literally means cured, and will keep for a considerable while without spoiling. It is interesting to note in passing that, among all the Western languages, Portuguese is the only one that has adopted the Arab word for the cured date as its own designation for the fruit. Whereas the English *date,* the French *datte,* the German *Dattel,* the Spanish *dátil,* and the Italian *dattilo* all have the same root, the Portuguese know the fruit as *tâmara.*

Size of individual dates varies considerably with the variety, and may range from one to three inches. Their shape is usually ovoid, although dates also may be roundish or egg-shaped and quite unlike the most commonly known kinds. Date varieties, of which there are many hundreds, are classed as soft, semidry, and dry; the dry dates are the so-called "bread dates" traditionally used by the Arabs as a convenient food, especially on trips because they keep their shape well, do not spoil easily, and provide satisfying nourishment. Although soft dates are also eaten in large quantities in Arab countries they usually are—or were in the past—pounded into a cohesive mass, or "cake," from which pieces could be cut off as needed. The sweeter, more sugary dates, which generally are the varieties preferred in the United States and other Western countries, where dates are considered a confection rather than a food, were usually eaten by the Arabs also only as a delicacy or dessert. Much of that crop was—and still is—sold for export in the Middle East and North Africa.

The ancient methods not only of growing and harvesting, but also of handling and storing dates in these countries remained basically unchanged until quite recently. They were described in graphic detail by Henry Simon in his "North African Diary" of 1913. Jotting down his impressions while observing and photographing the Arab date growers in their palm groves, and the merchants in the markets of the Algerian and Tunisian oases, he managed to assemble a vivid documentary not only of local methods of date cultivation, but of an entire way of life that had prevailed for centuries in those regions. This documentary is historically all the more valuable because modernization during the past few decades has brought great changes to this way of life.

One of the first signs that impressed itself upon the young American was the handling of the dates offered for sale in the local market places. "The Rhars dates I saw," he wrote, "were almost always kept in animal skins into which they had been trampled by the bare and unwashed feet of the proud proprietor and his helpers. Rhars is a very soft date that does not keep its shape at all well, and for that reason the merchants resorted to the trampling treatment, which in a matter of minutes turned thousands of dates into a horrible, dirty, sticky brown mess. The Arabs were not particular about the skins used as containers; mostly they were goat and sheep skins, but even cat and dog skins, and those of other small animals are acceptable. A skin of Rhars dates partially open is one of the most disgusting articles of food one can imagine, with the possible exception of those ghastly dainties made of the entrails of sheep, goats, and camels, which I saw offered on the meat markets of the desert."

Dry dates, Simon noted, received an entirely different treatment, even though that also could hardly be considered hygienic, at least by

Dry dates, the "bread of the desert," for sale in an oasis.

In this Algerian grove, Meshi Degla palms are seen on the l

Western standards. "The Meshi Degla, like other dry dates, were packed in large, heavy bags of camel hair. These sacks—which, by the way, were excellent examples of fine native craftsmanship, and frequently had truly beautiful color schemes—always came in pairs, because one was hung on each side of a camel. Every morning, these sacks were dumped, usually on the bare ground where camel and donkey manure was plentiful, and emptied—the same sacks, morning after morning, to be filled with the same dates every evening, minus only those that had been sold during the day."

...ars palms in the center, and the tall Deglet Noor on the right.

Those dates not used for home consumption but rather for export received special treatment. This was true especially of the Deglet Noor variety, the best of the dates grown in quantity in those parts of Algeria, and the one whose crop at that time, according to Simon's notes, was earmarked largely for export: "The Deglet Noor was sold in considerable quantities in that region. For shipment to the north, these semidry dates were packed in wooden boxes containing perhaps from twenty-five to fifty pounds, with the dates still on the twigs. Smaller quantities could be had by the fruit cluster."

Among the hundreds of varieties grown in different parts of North Africa and the Middle East, only a relatively few are cultivated in great quantities. All varieties have names which are often highly imaginative, such as the one called "Bridegroom's Fingers." In Algeria and Tunisia, the three principal varieties are the soft Rhars, or "Vigorous Grower," the dry "bread date" Meshi Degla, or "Purgative Seedling," and the semidry Deglet Noor, or *daqlet el nour* in Arabic, which is commonly interpreted as "Date of Light." Actually, the meaning of the name is uncertain; one story asserts that this variety was named after Noura, one of Mohammed's favorite wives, who was so blessed that a palm bearing better fruit than others grew on the spot where she once poured out some water. Arabs sometimes affectionately call the Deglet Noor the "Daughter of Light," since *daqlet* means "seedling" or "offspring."

The Deglet Noor is one of the most viable varieties, and the one grown predominantly in the United States, where today about 75 percent of the date harvest consists of this variety alone. Some of the most highly prized date varieties, such as the rare Menakher, the "Nose Date," were never sold for export, and the few existing palms were jealously guarded by their lucky owners, who steadfastly refused to part even with a few offshoots.

Morocco, Egypt, and the Persian Gulf countries such as Iraq all have their own distinct and favored varieties. Many of them are soft dates that are never sold for export. There is, for example, the soft Hayany date of Egypt; this particular date was named after a village. The Khalasa—"Quintessence"—is a soft date of Arabia. Soft varieties of Iraq and Syria include Barhee, named after the *barh*, the hot winds of the desert, Halawy, "The Sweet;" and Khadrawy, "The Verdant." Several of these soft dates are grown in small quantities in California.

Spectacular for its size, and renowned for its flavor, is the Medjool, or *mahjul*, of Morocco; rather incredibly, the name of this high-quality date means "The Unknown." The Medjool is a large, dark date that may grow to almost three inches long. Although classed as soft, it is much firmer than such varieties as the Barhee and the Khadrawy, and handles better. Before the First World War, the Medjool was one of Morocco's outstanding varieties; a few years later, however, the trees appeared doomed to extinction through a fatal plant disease that spread quickly and could not be checked. Moroccan officials therefore, in 1927, gave a few offshoots to representatives of the U.S. Department of Agriculture, who took them back to the United States for experimentation. The palms did very well in California, where they were planted, and today the Medjool, once close to extinction in its homeland, is alive and well in California, and is one of the choicest showcase varieties of date growers in the New World.

Depending upon the variety, a well-kept and well-tended date palm may yield from between 100 and 300 pounds of fruit annually. Given the right climate, the crucial factor in achieving good results is ample irrigation. A date palm can absorb incredible amounts of water, and justifies its nickname of "Friend of the Fountain." Some groves in the Persian Gulf region are irrigated every twelve hours by the tides, and thrive on it. They are hardy trees in many respects, and can be grown in either sandy or heavy soil, as long as there is sufficient drainage. Alkaline soils also do not bother them much, except if the salt content amounts to more than 1 percent. For a tree of desert regions, the palm is surprisingly resistant to cold, and is known to have survived subfreezing temperatures. Although ripening dates are sensitive to moisture in the

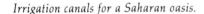

Irrigation canals for a Saharan oasis.

air, some varieties can tolerate more than others. The palms are generally free from diseases and parasites, although certain scale insects and fungi have been troublesome at times.

The work involved in caring for the date groves was always laborious and time-consuming, especially with the primitive tools and methods that until a few decades ago were universally employed in the palm's homelands. During a visit to the home of his native assistant in a small Algerian oasis, Henry Simon had an excellent opportunity to observe these Arab methods of planting, irrigating, and caring for their palms. He had already noted earlier that all well-kept gardens were cultivated each spring with the *messha*, a short-handled hoe, which had to be swung high above the head in order to drive the blade into the soil with sufficient force.

During an inspection tour of the date gardens belonging to his assistant's family, the American visitor was filled in on all the details of date palm growing, care, and cultivation. "It was to be seen in all the

Arab date grove owner at his hand-operated water pump.

A choice young date palm, planted below garden level as protection against hot winds and sun.

gardens," he wrote, "that careful and intelligent work had been put in; they were all very well tended. In the largest, the soil was very stony, and the planting of each young palm had required a hole eight feet deep by eight feet square in order to ensure it a chance to do well. Some of these holes seem to have been dug through solid sandstone. They were then filled with a mixture of good soil and manure to a depth of five feet, and then the offshoot was planted so that its own level was still three feet below the level of the garden. As each young palm grows, the remainder of the hole is gradually filled in. This method of planting, though undoubtedly the most costly of all I had seen, has the double advantage of protecting the young offshoot against both hot wind and sun, and ensuring the best use of water, which was applied by hand during the first few months."

The Arabs traditionally harvested their dates by simply removing the entire fruit cluster with the twigs, or rather strands, at the desired stage of ripening. Modern American growers, who are interested in getting dates at the peak of perfection, handpick many trees individually as the fruit ripens; the fruit clusters at that point have already been thinned out to permit the air to circulate, and thereby obtain more uniform quality fruit. That, of course, is a very laborious and costly process in this country because of the manual labor involved, and helps explain why choice American dates are somewhat more expensive—and at the same time better and more appetizing—than most of the imported fruit.

Although date palms may grow quite well in some areas with relatively high humidity, such as coastal regions, they can bear fruit successfully only in dry, hot climates with little or no rainfall at all. A much-quoted Arab proverb claims that the date palm has to have its feet in the water, and its head in the fire (some versions make that latter "hellfire") in order to flourish. For satisfactory fruit production, hot sun and dry air are just as necessary as liberal amounts of water for the root system. It seems strange, for a tree requiring so much water, to bear fruit which can be ruined even by small amounts of rain during certain ripening stages. Because dates are so easily damaged by rain, many American growers protect their fruit against an occasional rainfall by covering the clusters with a kind of cap, or cloak, consisting of paraffined paper bags. Open below, these bags permit the air to circulate freely, but shield the dates against both moisture and marauding birds. This invention of a California date grower has found wide acceptance; date palms wearing their bell-shaped paper cloaks around the fruit clusters are a common sight in California's date groves today.

Once the dates have reached the final ripening stage, they are no longer quite as susceptible to rain and humidity—unless, of course, there is any prolonged exposure to moisture, which in the regions where they are grown is highly unlikely. Some varieties, however, are so much more vulnerable that their cultivation becomes altogether uneconomical. The Rhars date, for instance, is no longer grown in the United States because it proved so extremely susceptible to moisture damage. All soft dates require extra care in handling and packing, a care that is supplied in the plants of California growers, and which makes possible the sale and shipping, throughout the United States, of high-quality soft dates in good shape and condition.

A most interesting trait of the individualistic date palm is the instability of any variety when grown from seeds. Through centuries of continuous experimentation, it has proved impossible to raise seedling palms whose fruit can be counted upon to have the same qualities as that of the parent palm. This does not mean that dates from a seedling palm are never good—they may be very good indeed, but they also may be quite inferior. Although seedlings are instrumental in getting new—and occasionally even desirable—varieties, the only way to reproduce an established and proven variety is through the offshoots. When cut off and planted, they will year after year produce fruit of exactly the same kind, size, texture, and flavor as that of the parent tree. Because of this characteristic, which renders impossible any "Johnny Appleseed" concept, the establishment in the United States of date groves capable of yielding fruit of known quality required large-scale importations of offshoots from the Middle East and North Africa. How and why North Americans initially became interested in this tree, and the subsequent efforts of early "date pioneers" to overcome all obstacles in pursuit of their goal to obtain and raise these palms in the Southwest, is one of the most romantic and intriguing adventure stories of American agricultural history.

Date palms planted around 1770 in San Diego's "Old Town" district.

4. Dates in the New World

From Spain to the Americas

Exact information about when and where the date palm was first introduced into the Western Hemisphere is difficult to obtain and even more difficult to verify. Just as scholars and historians are in doubt about the original home of this tree, as well as about the origins of its cultivation, so they also are not sure exactly when and where on American soil the first of these trees were planted.

What we do know for certain is that the Spaniards were responsible for introducing the palm quite soon after they claimed territory in newly discovered "West India." Of all the groups that were to settle in the Americas, only the Spaniards were truly familiar with the date palm; to later immigrants coming from England and Europe, it was an all-but-unknown, exotic tree. In Spain, however, date palms had long been "household trees," especially in the southeastern parts along the coastal region, where they had been introduced by the Phoenicians, evidently at the request of the Romans during the time when that part of the Iberian Peninsula was a Roman colony more than two thousand years ago. The

Phoenician concept of a sea wall reinforced by date palms.

palms were planted apparently not because of their value as fruit trees but rather as an aid in providing a dam against the disastrous periodical floodings of those regions by high tides. Establishing a living barrier of trees that could send down deep roots and also tolerate the salinity of the soil, and could fortify with their tall trunks the dams that were piled up against the tidal floods, evidently seemed to be the best way for handling that problem in those days. Fruit production was a secondary consideration; even if poor, any harvest would have been at least some additional gain from the trees whose primary function was to help protect the coastal regions from the ocean.

During the Moorish conquest of Spain, which began in the eighth century A.D. and lasted for seven hundred years, date palm cultivation became much more widespread, for to the Saracen invaders from Africa that tree was one of the most important of all fruit trees. The southeastern province of Alicante, whose capital, which is also a seaport, bears the same name, became the center of Spanish date cultivation, and today still numbers within its borders some 800,000 palms. Almost one-fourth of these are found in the huge *palmeral,* or date grove, near the town of Elche, the ancient Ilici in the times when those parts were a Roman colony. Located about twenty miles inland from the capital, Elche is known throughout the country as the "City of Palms" because many of its beautiful parks are distinguished by date trees that are offspring of those planted by the Moors.

The date groves at Elche yield a considerable annual harvest, which is one of the main economic assets of that region. Because of climatic and

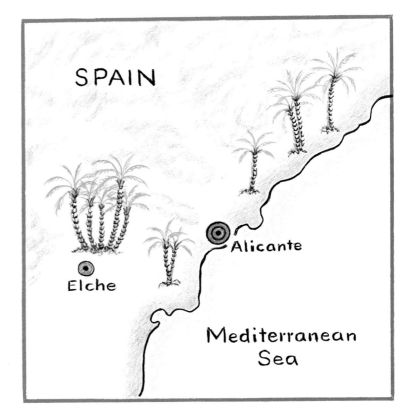

Spain's "Date Province" of Alicante.

SPAIN

Alicante

Elche

Mediterranean Sea

Symbolical gilded date palm used in the annual medieval mystery play at Elche.

other conditions, the dates are not top-quality and do not lend themselves well to preservation and export; instead, they are processed into a number of food products, including date wine and a kind of syrup known to this day throughout Spain as the famed *miel de Elche*, the "honey of Elche."

The town is even more famous, however, for the annual festival that draws tens of thousands of visitors in what is known as the Pilgrimage to Elche, where day-long celebrations, religious processions, and similar pageantry, including a medieval miracle play featuring the date palm as the central theme, awaits them.

It was perhaps inevitable that the Catholic Church would find a way to appreciate and include, in local religious ceremonies, the palm's importance to the region of Alicante. After all, the Church everywhere has shown a unique talent for assimilating, and fusing with church ritual, pre-Christian and secular concepts, traditions, and facts of local life. In this case, it was done by introducing the date palm into a religious play of the type known as *mysteries* during the Middle Ages. Mysteries in this connotation were ritualized dramas—usually accompanied by choral music and acted out by the local population—featuring certain incidents or passages from the Scriptures. The mystery play for which Elche was to become famous centers on the death and ascension of the virgin Mary. It begins as an angel, descending in a golden cloud shaped like a date palm, appears to the aged Mother of Christ and announces her imminent death. The angel carries a golden palm frond, which he hands to Mary with the explanation that from it a date palm will grow on her grave. Mary asks only for enough time to confer with the Apostles, a request that is granted her. She gives the palm frond to St. John, and he in turn passes it on to St. Peter. As she dies, her soul ascends to Heaven in the accompaniment of angels, and she is declared the Patroness of Elche. St. Peter baptizes repentant sinners by touching them with the palm frond, whereupon their sins are taken from them. The ancient religious symbolism of the date palm thus lives on in this play, adapted to the Christian setting.

In view of the tree's importance in their homeland, it is hardly surprising that the Spaniards sought to introduce it as soon as possible into their new colonies in the Western Hemisphere. The earliest available records inform us that there was an attempt at establishing a date plantation on Cuba in 1513, just a little over twenty years after Columbus had made his famous discovery. The attempt failed for the simple reason that its originator, a man by the name of Pánfilo de Narváez, was killed by hostile Indians. The murder of Narváez did not, however, go unavenged; retribution was exacted by one Rodrigo de Tamayo, who belonged to the group of soldiers led by the conquistador Velasquez. De Tamayo continued the work of Narváez in Cuba, and founded the settlement of Dátil, which of course is the Spanish word for date. At Dátil, he established both date and tobacco plantations. The climate was excellent for the latter, but not really suitable for good date harvests. Although the date palms did yield some fruit, it was not of good quality, and could be used only processed as syrup. Eventually, the date palms were all taken out, and today Cuba no longer has any at all; the settlement of Dátil, however, did survive as a suburb of the city Sur de Bayamo, and is a reminder of that early attempt of date palm cultivation in America.

Both the tobacco plant and the palm were brought from Cuba to Mexico when expeditionary forces led by the conquistador Hernando Cortez invaded that country. The "Conqueror of Mexico" liked to make a point of claiming that he was the first to introduce into the new colony not only the date and eucalyptus trees, but also the tobacco and *yuca* plants. Although the well-known desert yuccas of the American Southwest which belong to the lily family were named after the original *yuca*, they are not closely related. Often called cassawa, the *yuca* is a tropical plant whose rootstocks yield a nutritious starch which, in the granular

Location of the New World's first date plantation in Cuba.

form known as tapioca, is used for soups, puddings, and the like.

Unlike Cuba, Mexico has many arid, desert areas which lend themselves well to date cultivation. Date palms still exist today in some parts of the country, even though there never was any really large-scale cultivation effort, and their economic importance has been negligible. Once the trees were established in Mexico, it was only natural that, as the Spanish claimed more and more territory to the north along both the Gulf of Mexico and Pacific coasts, they included palm plantings among their other agricultural projects in the new colonies.

The most advanced outposts of the Spanish empire were the missions; the intrepid padres who manned them were introducing not merely a new faith, but also new skills, customs, and ways of life in the regions where they settled. Some of these novel ways were destructive, others beneficial. Among the latter must be counted many of the new domestic animals and food plants they introduced, as well as the considerable agricultural knowledge they brought with them. The missions therefore soon became the centers of Spanish influence in the new lands.

Although the available records yield no information on this point, it seems quite probable that date plantations were attempted by missionaries in some parts of Florida during the early days of that region's conquest by the Spanish. As in Cuba, such plantations would have been destined to fail because of the high humidity of the Florida climate; if they did exist, no trace—not even a name—remained of these groves. Matters were different along the Pacific coast, however, for we do know that date palms—all of them raised from seeds instead of from offshoots—were quite common around many of the missions of the Jesuit and Franciscan orders in southern California. These palms were probably not planted primarily for their fruit—which the Spaniards must have known to be sensitive to the humid coastal climate—but rather for the stately good looks of this palm. The date would have been only one of the many ornamental as well as food plants with which the horticulturists among the good padres experimented on the land of the newly acquired American colonies.

As far as can be estimated today, the date seeds planted around the missions in San Diego came from Baja California. Records show that the ship *San Carlos,* loaded with supplies for Spanish settlements in Alta California, sailed from La Paz in Mexico in 1769; the ship's provisions included two *tanates* of dates, and one of her first ports of call was San Diego. A *tanate* was a basket, or hamper, holding approximately seventy-five pounds when filled, and used widely by the Spanish for produce such as dates. Further confirmation for the Mexican origin of

Mission San Diego de Alcalá in a drawing made in 1854

the old date trees of San Diego can be found in their close resemblance
to the variety grown in the La Paz region of Mexico.

Some of the date palms planted on mission grounds in the late
eighteenth and early nineteenth centuries flourished, and a few—such as
those at the "Mother of Missions," San Diego de Alcalá, founded in 1769
by Fra Junípero Serra—survived until after World War II, when they
finally had to be taken down because they were dying. The oldest of
these trees thus attained an age of almost two hundred years.

There is no evidence that the Spanish attempted to establish large date groves anywhere along the California coast, but with the introduction of the palm, a great number of seeds were planted during the settlement of the interior regions, and especially during the years of the gold rush following the discovery of that coveted metal in California in 1849. Some of these seedling palms later were to play a decisive role in spurring efforts to try large-scale cultivation.

It seems that interest in this tree developed within a few decades after the United States had become an independent republic; some enterprising Americans were sufficiently intrigued to try raising dates in Florida after that state had been acquired from the Spanish. Unfortunately, the knowledge of those early date pioneers did not match their enthusiasm. When it became apparent that climatic conditions in Florida were unsuitable for date cultivation, they discontinued their attempts, and shortly thereafter all thoughts of raising dates commercially in the Southeast were abandoned.

One of the earliest American publications discussing the desirability of date palm cultivation in the United States is an article that

By 1890, both the mission and its palms were in a state of utter neglect.

appeared in the *American Monthly* magazine in 1818. Written by one S. L. Mitchill, about whom very little else is known, it bears the rather lengthy title "An Encouragement to the Introduction of the Date-Bearing Palm into the United States." After discussing some of the great advantages of cultivating the "date-bearing palm," Mitchill mentions that a man by the name of Henry Austin, who apparently was an importer of dates and other fruit, had also imported a few offshoots from the Persian Gulf region because he had heard that seedlings were unreliable. No one knows, however, what became of these offshoots. If they were indeed high-quality palms—which appears very unlikely under the circumstances—they probably did not survive because their owner knew little or nothing about the trees' requirements and did not treat them properly.

By the early 1830s, all efforts to raise date palms commercially in the coastal regions of the United States had failed, and interest in establishing plantations had all but disappeared. It was to remain negligible until almost half a century later. At that time, a few palms grown from seeds in some of the hot, dry interior regions of the West began to yield fruit, and interest in the possibilities of date cultivation was immediately revived. The idea of establishing commercially viable groves took hold among a number of plant specialists intrigued by what they had learned about this tree. Few of them lived to see the realization of that goal, which was to require another forty years, countless experiments, and persistent efforts by enterprising and adventurous pioneers.

So far as we know, 1877 was the year the first dates were harvested in the West. They came from palms planted as seeds by rancher J. R. Wolfskill in the Sacramento Valley exactly twenty years before. In the early 1880s, seedling palms near Yuma in Arizona, planted during the Civil War years, began to produce fruit; they were followed in the next decade by several others—all of them raised from seeds—in parts of Arizona and California. At that point, experts of the U.S. Department of Agriculture—in those days a small agency staffed mainly not by bureaucrats but by imaginative and enterprising men—began to turn their attention to the successful production of experimental date crops in the arid regions of several southwestern states. Commercial growers were not involved in those early efforts; although a few small offshoots had been sent in 1876 to California from Egypt, these palms eventually died through the neglect of the owners, who either did not care or know enough to give them proper treatment.

It appears that the first date palms grown from offshoots that survived for any length of time were imported by employees of the U.S. Agriculture Department in 1890. That first lot was a small quantity of some seventy-five palms, most of which came from Egypt, although a few had been purchased in Algeria and Arabia. These palms went to various parts of the Southwest, including New Mexico, Arizona, and California. Unfortunately, it was not a very successful undertaking; the offshoots had been obtained through the help of American consular officials, who naturally could not be expected to know anything about palms and had to rely on local help. It appears that most, if not all of them, were very inferior plants, and quite probably seedlings doctored to look like offshoots. Although some of these trees survived until after World War II, all of them were subsequently removed.

The first really successful importation of name variety offshoots was made in 1900. These trees—mostly of the Deglet Noor variety—came from Algeria, and marked the beginning of serious attempts at professional date cultivation. That first import operation was supervised by Dr. Walter T. Swingle of the Department of Agriculture. Most of the approximately four hundred offshoots were sent to Arizona, and were planted near Tempe as a research project.

Up to that time, and continuing through the next few years, practically all the experimental plantings were made by—or under the supervision of—the U.S. Department of Agriculture. This research included the study and development of the best methods for cultivating date palms in the United States as well as selection of the varieties best suited to American climatic and soil conditions. Prior to the turn of the century, private growers were not involved at all.

Palms grown at the experimental stations came from many different parts of the Middle East and North Africa. Testing and research continued long after large commercial groves had been established in the United States. The last such importation by the Agriculture Department, of several new varieties from Iraq, was made in 1929. Today, American know-how on the subject of date cultivation is acknowledged by those Middle Eastern growers, who ask for American consultants when they have difficulties with their groves.

In the years before the First World War, the experiments in date cultivation conducted by the Department of Agriculture had attracted the attention of some enterprising individuals. Although they all wanted to make date cultivation economically feasible, it was the romantic history of the tree rather than just its money-making possibilities that inspired some of the most colorful among the early pioneers. All of them, however, believed that large-scale cultivation of date palms could be successful in certain parts of the American Southwest, where climate, soil, and plant life closely resemble those of the arid regions of the Middle East and North Africa where most of the Old World date groves are concentrated.

The earliest of the private importations was made by Bernard G. Johnson, one of the most enterprising of the early date cultivation pioneers. Johnson went to Algeria in 1903, and returned with 129 offshoots, mostly of the reliable Deglet Noor variety, which he planted on a parcel of land he owned near Mecca in California's Coachella Valley, not far from Indio, the present "date capital" of the United States. The Deglet Noor date was then—and still is today—considered commercially the most viable of all the varieties, and the one with the largest average fruit yield.

Johnson imported another small lot of offshoots in 1907. These palms thrived; there could no longer be any doubt that the climatic and soil conditions of the Coachella Valley region were ideally suited for growing healthy palms and obtaining successful harvests. Large-scale importations were, of course, mainly a matter of money. For one thing, the offshoots could not be simply ordered by mail, sight unseen. The only way of getting high-quality palms was to send someone with enough knowledge to select them to the Middle East or North Africa. Such emissaries had to possess the diplomatic skill necessary for obtaining government export permits, as well as for striking the best possible bargains with frequently reluctant native date grove owners. At the other end, in the United States, the sites for the new plantations had to be in arid regions, yet amply supplied with the large amounts of water the palms required. At that time, this water could be obtained only through the costly process of digging or drilling deep wells in the desert.

It was therefore only after Johnson had inherited some money that he was able to finance a major buying trip to Algeria in 1912, from where

Marker noting the California date industry's birthplace in Mecca.

he brought back 3,000 Deglet Noor offshoots, the largest date palm importation ever made up to that time. However, his hopes of finding buyers for these palms in California were disappointed, because in the meantime rival growers had made their own plans. Johnson therefore sent his palms to Yuma, where he planted them on his own land.

In 1912, Arizona thus actually had a greater number of date palms than California. This, however, was to change almost overnight because the "date fever" had really caught on. Large-scale importations by private companies began in earnest early in 1913, and continued throughout that year, thereby establishing the Coachella Valley as the center of American date cultivation. Within that one year, more off-shoots were imported than in all the earlier years combined, and almost as many as the total of the later imports between 1914 and 1922, when the last large importation was made from Egypt.

In order to understand what such importations prior to World War I involved, the difficulties of traveling in many of the regions where top-quality offshoots could be obtained have to be appreciated. Especially in such French colonies as Tunisia and Algeria, Western travelers who had to leave the beaten tracks and venture into the Sahara Desert incurred many risks and dangers. The means of travel in much of North Africa were still extremely primitive; large areas of the interior had no railroad, and although some stagecoach lines existed, many places could be reached only by camel, horse, or mule in exhausting and uncomfortable journeys across the desert. Transportation of the offshoots to the coast was possible by camel caravan only. The purchasing of the young date palms in the desert oases added another risk because of the large sums of cash the buyer had to carry among people who often committed murders for no more than a few dollars. Matters were not exactly helped by the frequent local native uprisings in what then still were French colonial lands; such upheavals made travel hazardous for any stranger caught in the middle. Only men of pioneering spirit, cool nerve, and stamina could successfully handle the difficulties attending these date palm buying expeditions.

Really large-scale importations into California began in 1913, and

Land cleared and ready in 1912 for the Coachella Valley's first large commercial date plantation.

the largest in that year—or in any one year, for that matter—had been planned by a company known as the West India Gardens, with headquarters in Altadena, California, but with a branch office in Thermal in the Coachella Valley. That office was headed by Paul Popenoe, a local rancher who was very much interested in the date cultivation project. The West India Gardens owned large tracts of land east of Thermal near the Mecca Hills, and had designated that land as the new home for the thousands of date offshoots they planned to import from the Persian Gulf region and from several North African countries.

Adjacent to the land earmarked for that nursery was the property of Henry Simon. The well which he had drilled on his land in 1912, and which was capable of supplying the large amounts of irrigation water needed for the projected plantation, was located on the northwest corner of his property very close to where it bordered on the West India Gardens tract. The necessary irrigation apparatus and distributing system was therefore easy to map out and control. Under the circumstances, it was only natural that Simon and his neighbors of the West India Gardens pooled their forces for the benefit of the planned date cultivation project.

The company's plans called for an importation, in 1913, of a total of approximately 15,000 palm offshoots of different varieties from the Middle East and North Africa. While Paul Popenoe planned to travel to Iraq, the purchasing of the crucially important Deglet Noor offshoots, plus small experimental lots of other Algerian and Tunisian varieties, was entrusted to Henry Simon. He readily agreed to undertake the venture; in addition to his knowledge of French and Arabic, he also had the necessary expert knowledge to judge the offshoots and select good, vigorous specimens. This was no task for amateurs, for the past enterprising Arabs were known to have not only sold male instead of female offshoots to unsuspecting Americans, but even to have occasionally doctored seedlings to make them look like offshoots.

The adventurous and romantic aspects of the date importations were largely—and quite naturally—overshadowed at the time by the important details concerning the successful cultivation of the imported

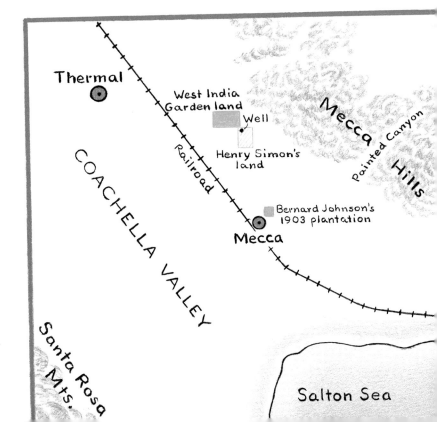

Mecca and surrounding area with locations of early date groves.

meskhah, manure.
All carried on the
back of burros and
men.

pollinating. How
the fruit grows. Age
and size of palms.
Palms without water
for 5 years.

Phus. The "most best"
of all dates.
& the two Phus and
Deglet Noor."

Men from the Souf
working in front of his
store.

The Date Palm. What it is, and what it is to the
Arab. The foundation of Desert Arab Life. All desert
Arabs not XXX Date Arabs. The Desert Arab and the
"Arab of the Mountains", the Bedouin. The camel, the
date-palm, and the Arab. Kinds of dates, quantities
harvested, selling the Deglet Noor crop. What all
the palm gives to the Arab. Tatai Ben Mustapha's talk.
Food for man and XXXXX beast, fibre for ropes and mats,
leaf-material for baskets and hats, wood for building
and dam-work, lagmi when the palm is dying, sticks
for tents and for walking, food in the crown, fire
wood XXXX from the leaves.

The burro and the Arab child.

The date and what it is and how it is marketed.
Disposing of the Deglet XXX Noor crop. The dry dates,
the bread of the desert. Other foods besides the date.
The meat market. Camel meat. Dainties for the Arab
palate. Hygienic conditions of the markets. Where
you are neverX sure what you eat. Taking a meal with
the Arab.

Tatai Ben Mustapha. Cigarettes and cahwa sucra
ala barra. Making the contract. The store of Tatai.
The stories of Tatai. Going out for prayer in the
public market. Locking the store. The rise of Tatai
to prosperity. What Tatai bought, - a date garden,
a donkey, a wife, another date garden, another wife,
another burro.

The caravans I sent off. Beating them home by
stage. The fondouk at Biskra. Offshoots entering. The
beautiful camel-driver. Trimming and packing the
offshoots.

The camel. Costs nothing to
maintain. Burden of a camel.
Customs. Cruelty. Basham or
letter to recipient. The
camel of burden and the
mehari. The mehari by the
side of the stage. The annual
mehari race. How caravans
appear. Hobbling camels. Ben
Ali cracks a joke.

Buying palms in the
market. Prices.
Carrying them to
the store. What I
paid my men. The
one-eyed scoundrel
Ben Touche.

Protecting the date garden from the evil eye.
Aderrahman endorses the evil eye. Homely and cheerful
ideas concerning the laws of biology. How a male offshoot is converted into a female. It always works -
only sometimes not. The Arabs idea of seedling dates.
Degla. They never amount to anything- only sometimes
yes they do.

A page of Henry Simon's original North African travel journal.

young palms. Unfortunately also, most of the early pioneers, including the intrepid Bernard Johnson, who braved considerable risks in some of his early journeys, left few if any records of their travels. Yet these romantic and adventurous aspects are an intriguing facet of the date palm cultivation project, which in turn is a small but fascinating chapter in America's colorful history of agricultural enterprise and development.

Fortunately, the detailed notes jotted down by Henry Simon in the oases of the Sahara were preserved, as were the hundreds of film negatives of the pictures he took during his more than five months of crisscrossing Algeria and Tunisia by stagecoach and on horseback in quest for offshoots. Guided by this diary, which is the only documented record of that enterprise, we can accompany him on his search for the *djebbar*, as the Arabs call the offshoots, and find out what such journeys were like in those days.

Henry Simon with an Arab grove owner and a load of offshoots.

5.   Date Palm Safaris

Adventures in the Sahara

Henry simon arrived in Algeria in February of 1913, and in the subsequent five months traveled widely in that country as well as in neighboring Tunisia, covering a total of several thousand miles, and visiting oases far south in the Sahara. From the very first, he decided to keep a daily record of his journeys. "Despite the work and worry involved in selecting and purchasing thousands of date palms for exportation to California—at that time the largest exportation of its kind in the history of Africa," he wrote later, "I still managed to find time for my travel journal and my camera. Thus my notes on that trip were documented by some five hundred photographs, a copious and detailed pictorial record designed to keep my memory fresh."

After establishing headquarters at Biskra, an Algerian oasis only about 150 miles from the coast, and even then a favorite French tourist town, Simon proceeded as first order of business to recruit a native assistant. Selection of the right kind of man was of paramount importance to the success of his mission. When he found one who throughout

Ben Ali makes a point.

their subsequent association proved to be not only honest and reliable but thoroughly devoted to him, it was due as much to good fortune as to good judgment of character. The full and imposing name of this Arab was Abderahman Ben Ali Ben Haouffef, but he appears simply as Ben Ali in Simon's diary. Accompanied by this assistant, and armed with a

Street scene in the oasis town of Biskra.

French government permit for the purchase and export—for a specified area only—of 2,000 Deglet Noor offshoots, he was ready for his first excursion into the Sahara. His goal was the large oasis of Touggourt, which is located about 150 miles south of Biskra, where he hoped to find quality offshoots at a reasonable price.

Simon's success in quickly obtaining the necessary government permits to buy palms in both Algeria and Tunisia, which originally covered some 3,000 offshoots, was acknowledged by the home office in a congratulatory telegram; at the same time, he was urged to try for additional permits totaling another 3,000 palms, for a grand total of 6,000 offshoots the company wanted from that region.

While waiting in Biskra for further details from California, the young American spent some time acquainting himself with the modes and manners of Arab life in the oases, and with studying Arab methods of date cultivation. A week later, he and Ben Ali took the stage to Touggourt to scout around for the first lot of offshoots.

The stage ride through the desert was uncomfortable and cold during the night, but otherwise uneventful. To the American, the surrounding countryside and its plant life looked surprisingly like that of the low-lying salt regions of the Colorado Desert which he knew so

Date market in Biskra.

well. "There was," he noted in his diary, "practically no difference the eye could perceive, though learned scientists undoubtedly could have told me that—owing to the different formation of the *guzuzicus*—the character of the Saharan plants is different, and that therefore what looks, feels, and behaves like *Hinkadihankicus multiplex* is really a variety of *Hunkadibunkicus simplicissimus.*"

The first fourteen hours of travel were interrupted only once for a change of horses, five of which pulled the stage during that part of the journey. Then came the halfway station, at which the drivers and passengers got off for a night's rest and some abominable food consisting mainly of tough meat from an ancient sheep and some vile coffee. The only good part of the meal were the dates offered for dessert. At two o'clock the next morning, the drowsy passengers were awakened to begin the second, somewhat longer but much smoother part of their journey, with now only three horses easily pulling the coach. Sixteen hours later, they reached Touggourt, one of the largest and most beautiful of the Algerian oases.

Shortly before arriving in Touggourt, the coach passed through several small oases with good-sized date palm groves. What struck the

The diligence, *or stagecoach, in Algeria in 1913.*

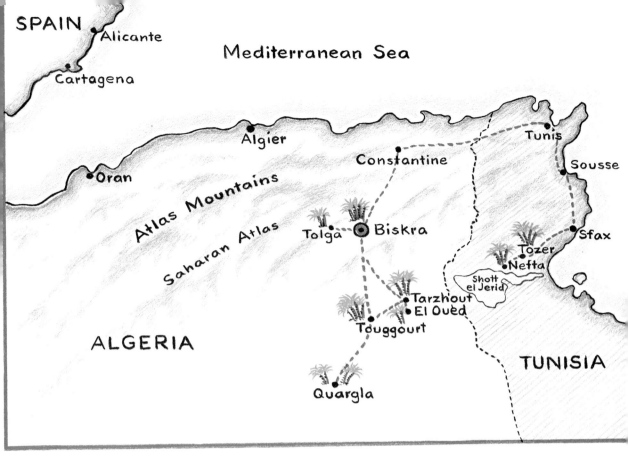

Henry Simon's travel routes in the Sahara.

American observer as peculiar about some of these groves was the sickly appearance of the Deglet Noor, which seemed to be suffering considerably, whereas other varieties growing right next to them were doing well. Simon put that down—correctly, as it turned out—to an overabundance of salt in the soil, which the Deglet Noor palms, whose root system is relatively shallow, cannot stand as well as those varieties capable of sending their roots down below the layers of salt, which accumulates mainly in the topmost few feet of soil.

Immediately after arriving in Touggourt, Ben Ali set out to locate the owner of a sufficiently large grove of Deglet Noor palms who would be willing to sell to a foreigner. He returned that same day with the good news that he had found a merchant by the name of Tatai Ben Mustapha, who owned such a grove and had tentatively agreed to furnish the entire

Taking coffee with wealthy Arab date grove owner Tatai Ben Mustapha.

2,000 offshoots covered by the French government permit for the Touggourt region. However, he first wanted to meet the *Amerikany* before closing the deal. The arrival of an American in the oasis was understandably a sensation; everything about him from his soft, Indian-style leather leggings to his camera was an object of interest.

The meeting with Tatai Ben Mustapha was highly satisfactory to both parties; Simon found the Arab agreeable and his prices moderate.

Soon the contract for the purchase of the first batch of offshoots was made and signed before the official military notary with the help of countless little cups of very strong coffee for everyone. "It should be mentioned in passing," Henry Simon noted in his diary, "that *cahwa* (coffee) forms an indispensable part of business in those regions, for coffee has to be drunk or business cannot thrive." So much coffee did he consume during that first transaction, misled by the smallness of the cups, that he became quite ill afterwards. For hours he lay stretched out on his bed in the hotel, unable to get up, with his heart pounding away at a mad pace and the blood thundering in his ears. From that day on, he always carefully watched the amount of *cahwa* he consumed on any one occasion.

The Arab population of Touggourt was most friendly and eager to help the young American in any way possible; his command of both French and Arabic, as well as his natural gift for striking the right note, smoothed over obstacles and opened doors that otherwise might have remained closed to a Westerner and non-Moslem. In addition, his camera intrigued young and old in the oasis; everyone wanted to have a

Snake charmer entertains rapt audience in a Saharan oasis.

picture of themselves and their relatives, and so he soon became the unofficial town photographer of Touggourt, and an immensely popular personality.

During his stay in the oasis, Simon was invited to inspect many of the palm groves, where he made notes on local methods of cultivation and irrigation. He later found that water application varied from one locality to another according to the soil. Wherever the soil was heavy, water was applied by means of ditches and basins; in locations with light sandy soil, the entire grove was flooded. This latter method was necessitated not so much by the date palms but by the presence of secondary crops such as figs, apricots, bananas, and oranges, which grew scattered among the irregularly planted date palms. This way of planting increased the beauty and visual appeal of the groves but also the difficulties of cultivating the palms. Simon correctly estimated then that in California, irrigation with just a single ditch along each side of a straight row of palms would be quite sufficient, and would not only save water, but also discourage the palms' undesirable tendency to form too many surface roots.

Arabs washing clothes and bathing in a grove's irrigation channels.

Irrigation basins in a date grove with hard soil.

Touggourt was only the first station in Simon's quest for offshoots; other trips took him even farther into the Algerian Sahara as well as to such large and picturesque Tunisian oases as Tozer and Nefta.

In Tunisia, the visiting American was one of the few non-French foreigners favored with an invitation to the fabled home of M. Jean Martel, the wealthiest and most influential colonist of that region. The house and gardens had an Arabian Nights quality that left Western visitors breathless. The lush vegetation of the garden with its exotic blossoms, interspersed by sparkling fountains and pools, and lavishly illuminated at night, provided a perfect setting for the large and luxuriously appointed house. The sumptuous dinner served to the guests featured the choicest and most delectable foods and wines. Much more important to the American were the personal letters of introduction to French officials he received from his host, for such recommendations made his job of negotiating the necessary purchase and export permits a good deal easier.

The oasis of Tozer is one of the largest and most beautiful in that part of Africa, and we have Simon's description of what he saw as he explored it. "Wandering along the roads that led through that great forest of palm trees," he wrote, "had almost a fairy-tale quality. There

Road through a large Algerian oasis.

The oasis of Nefta; the town is visible on the hills in the background.

was the rich bright green of the secondary crops that covered the floor shining up, flecked with gold by the sunlight that played on them as it filtered through the feathery tops of the palms. There were the buds of figs and other trees just coming out, the snowy blossoms of the apricots, the golden fruit of the oranges, and the broad gleaming foliage of the bananas showing through the gray and green of the palm trees. In many places grape vines hung draped like garlands from palm to palm, and every now and then one emerged from that luscious jungle to come upon one of the small streams that wind their way through the oasis, and without which there would be none."

Tozer was less satisfactory in regard to the special purpose that had brought the American there, for despite a large number of Deglet Noor palms at that oasis, there were not large enough numbers of offshoots to

Nefta displays typical oasis town's beehive of interlocking buildings.

be had at a reasonable price to warrant the time, trouble, and expense involved in a separate camel transport to the coast. Simon therefore decided to go on to Nefta, an out-of-the-way oasis rarely ever visited by Europeans in those days, whose entire non-Arab population numbered fewer than a dozen. Located on the borders of the great Chott Djerid, whose lifeless wastelands stretch far to the southwest, Nefta's oasis winds partly around, and partly through, the town itself like a broad green belt. A *chott,* it should be added, is a great natural evaporating pan where much of the underground waters of the desert rise, and there leave on the surface an accumulation of salt.

As before in the Algerian Sahara, the American visitor was deeply impressed by this desert's brooding yet strangely moving countenance. "There was a majestic desolation," he wrote, "a threatening vastness, a sad yet eloquent stillness about the desert as it shimmered in the brilliant sunlight that could not fail to impress. It was also a monotonous spectacle, yet of that I did not think, so mighty and solemn and awesome was it."

During his inspection of Nefta's palm groves, Simon found that he could have purchased a few hundred offshoots, but on closer examination found them to be long, thin, and spindly, a type he considered quite undesirable for exportation. He therefore decided to return to his headquarters in Algeria, and try some other oases there.

The return from Nefta to Tozer was distinguished by one of those little incidents that made travel in those parts so unusual and interesting. With about one-third of the way to go, Simon dismounted from his mule to stretch his legs and at the same time give the animal a breather. Instead of appreciating this kindness, the ungrateful beast refused to let its rider mount again, taking advantage of the fact that the young American, used to his Western ponies that followed him around like dogs, had let go of the reins. For one hot and exhausting hour Simon played tag with what he describes in his diary as "a jackass evidently possessed of an overabundance of grey matter." Finally, after a flying tackle had succeeded only in landing him in the sand embracing the saddle, whose half-rotted leather had given way, he conceded defeat and walked the remaining five miles back to Tozer, leaving a victorious jackass behind him in the desert.

On his return to Biskra, he found a telegram from Touggourt urging

Street in a Saharan oasis town.

him to get there for the cutting of most of the initial lot of 2,000 Deglet
Noor offshoots, which he wanted to supervise personally. He first had to
wait for a reply to his report to the West India Gardens home office, in
which he had requested more detailed information concerning additional purchases. During that waiting period, he decided to accept the
invitation of his trusted Ben Ali and visit the latter's family home at
Tolga, a small but pretty oasis located only twenty-five miles west of
Biskra, and at the same time explore the possibility of buying small
experimental lots of date varieties grown in that district.

Accepting Ben Ali's invitation to his family's home for dinner not
only afforded the young American an interesting chance for firsthand
observation of orthodox Mohammedan family life; it also further
cemented his relationship with Ben Ali, who felt greatly honored that his
esteemed *Amerikany* employer had agreed to be a guest in his house.

All went exceedingly well with that visit. Ben Ali's aged parents

Ben Ali in his family's well-tended date garden.

The great well of Tolga, in 1913 the largest artesian well in the world.

called down Allah's blessings upon the *Amerikany* who had been so good to their son. Following the inevitable *cahwa*, Ben Ali and his brother Mohammed took their visitor on an inspection tour of their small but well-kept gardens. He also had to admire the huge newly finished artesian well of Tolga—said at that time to be the largest in the world—and meet and photograph the old Arab who was the proud owner of the well, which he had kept on digging in the face of predictions of failure, and which had turned out to be a veritable gold mine. Toward evening, the Haouffef brothers returned home with their visitor. The women had been busy cooking all day, and dinner was ready to be served.

Knowing that Arab women in those days could never join the men at a meal, Simon had expected the repast to be an all-male affair. However, it did come as something of a shock to him when he found out that it also was going to be a one-man affair, at which he, the guest, was the only one supposed to eat anything at all, while the others sat around watching. Although he had not anticipated anything like it, and found

Henry Simon as guest in Ben Ali's home. Although the latter was an orthodox Moslem, he permitted the women of the house to appear unveiled before his American employer in a rare gesture of esteem.

the idea rather appalling, the young American was careful not to show any dismay or discomfort so as not to upset his hosts, and decided to make the best of the situation. "Out came a huge dish," he noted in his diary, "filled with a sort of half-soup, half-stew which consisted of cooked barley groats—the famous *kuskus* of the Arabs—with pieces of meat. I ate plentifully of it to show, Arab fashion, that I honored their meal, and also because I was fairly sure that not very much else would follow. But there I was sorely mistaken, for soon there followed a dish of meat, and of that I had to eat plentifully; and then followed roast chicken, and of that I had to eat plentifully; and within minutes followed two kinds of freshly baked Arab bread, one soft, one hard, and of both I had to eat. Finally, when I felt certain that another mouthful would surely choke me, there came sweet dates, and of those I had to eat also.

During the entire meal I had the company of both Ben Ali and Mohammed, his brother, but they did not eat with me; they merely sat around watching with the proverbial Argus eyes that I ate enough. Never should it be said that the white man did not get enough to eat in their home. It was never said."

The evening was a great success; when Ben Ali permitted his employer to take pictures of the entire family, including the unveiled women—an unheard-of sign of esteem and trust for an orthodox Moslem—Simon knew he had really scored. Later, after he had somewhat recovered from his mammoth meal, the two Haouffef brothers, both armed with rifles, accompanied their honored guest on the ten-minute walk back to his hotel. This armed guard was not altogether just a matter of pride, for murder was a commonplace occurrence in those parts; during the American's first week in Biskra, a town of barely eight thousand inhabitants, six murders were committed in broad daylight. In addition, there was then considerable internal strife through much of Algeria, which had prompted some French army officers to tell the young foreigner quite bluntly that he would be risking his life by adhering to the itinerary he had mapped out. "I was warned by experienced French officers," he wrote, "that if I persisted in traveling to some of the places I planned to visit, I was more than likely to come back minus my head. Despite these cheerful predictions, I kept to my plans, nothing untoward happened, and I returned unscathed. *Yektubah Allah*—God hath written it—say the fatalistic Arabs. All the same, I knew perfectly well that my life would not have been worth the proverbial thin dime if word had ever leaked out that I was carrying hundreds of dollars worth of cash in a money belt under my shirt."

Returning from Tolga to Biskra, Simon left Ben Ali behind and directed him to look for small lots of offshoots of several varieties he wanted to purchase from the Ziban district west of Biskra, while he himself again traveled to Touggourt to supervise the cutting of the Deglet Noor offshoots he had bought there.

The first order of business after his arrival in the oasis was the renting of a suitable storeroom for the *djebbar*, as well as provisions for

safeguarding the young palms during storage. Simon selected as watch-
man a strong young Negro with the euphonious name of Mohammed
Kidoona, who from that day on proudly worked, lived, and slept in the
"Amerikany's House of Djebbar," as the large shed locally became
known. The offshoots were transferred to that storeroom immediately
after cutting, usually in batches of a few hundred at a time, and every
single one was checked out by Simon, who rejected any that were either
too small, or not cut properly. Whenever he accepted an offshoot, he
wrote his name in full on one of the leafstalks before it was put away in
the storeroom, a simple stratagem which prevented any possibility of
substitution at a later date.

Later, after the faithful Ben Ali had returned from his trip to the
Ziban district, he took over much of the job of checking out the
djebbar—always subject, of course, to his employer's final approval. After
about half of the offshoots had been inspected, Simon engaged a caravan
of some forty camels, and personally supervised the wrapping and
loading of the young palms for their week-long trip through the desert to
Biskra. Each camel carried a load of approximately four hundred
pounds, which had to be evenly distributed on both sides of the animal's
back. Despite all his efforts, the young American could not at that time

Signing each offshoot to prevent substitution.

Brought fwd:	(6969.00) 6968.95
April 17	
Telegr. W. Pop.	-. 90.
Telegr. Mabon	1. 20.
Tel. W. Pop	-. 95.
Fel. Ch. Petite Vit.	-. 90.
19.	
Tel. W. Pop.	1. 25.
4 men at 1.50.	6. — .
1 do at 2. — .	2. — .
1 do at 2. — .	2. — .
Tel. Wils. Pop	2. 15.
Over Time pay	2. —
Coffee for 8	1. — .
Pd for 50 sacks	27. 50
Pd. watchman	1. — .
Lif	9. — .
Thread & Needles	3. — .
Lif	-. 75.
Papers for Bushman	1. 20.
Pay for 4 men	12. — .
Over Time and for hire of pails	2. — .
	7045.75
	7045.90

Le 16 avril 13.
Reçu de M. Simon
soixante cinq francs
(frs 65. —) que j'ai prêté
à Abdurrahman et
cinquante six francs 50 cent
(56.50) pour 56½ Ros
de cordes.

[Arabic script signature lines]

Le 24 avril 13.
Reçu de M. Simon frs. 100.
(cent francs) d'avance sur les
1500 francs qui resteront à
payer quand la livraison
des djebbar est complété.

[Arabic script signature lines]

Pages in Simon's expense account notebook with Ben Ali's signatures.

purchase sacks for the transportation of the offshoots, and had to be satisfied with the Arabs' usual method of wrapping the *djebbar* for transport, which consisted of a thick packing of leaf material, especially through the center where the roots and the terminal buds were located. For his later shipments, some of which were purchased in oases even further south in the Sahara, he always used sacks to wrap the offshoots securely.

The weather during the time of that first shipment—it was early

Loading the camels with offshoots.

March—fortunately was cool. After everything had been properly packed and prepared, Simon ordered *cahwa* for all, and drank it in the company of his laborers and the caravan leaders—a gesture which, as he noted in his diary, "greatly pleased everyone." He continued to describe what else had to be done before the caravan could go on its way: "The two leaders of the caravan and I appeared before the *bashamar*, the official transportation agent, who drew up a contract in which all conditions were set forth, and whereby the camel drivers were bound to deliver within six days at Biskra a letter addressed to *Henry Simoon, Amerikany,* and also 1,100 *djebbar* in good condition, and each with my name signed in full on a leafstalk; and that nothing but an act of Allah was to excuse them from doing so; and that, if they failed to deliver the goods or if any were missing, I would have the right to take their camels in exchange—a cheerful idea, my sitting around at Biskra with forty blasted camels on my hands."

The caravan duly departed; it was the first of many that were to follow in subsequent weeks and months from different parts of the country. Simon left for Biskra the very next day to locate and rent a suitable storage house, or *fondouk*, for the offshoots before the caravan arrived, and was lucky to find one large enough to accommodate several thousand offshoots, each of which averaged a length of six feet, and a weight of between fifteen and twenty-five pounds.

The caravan arrived punctually on the sixth day, and was directed to the *fondouk*, where the camels lay down and were relieved of their burdens. After all the offshoots had been counted and found in order, the American paid the caravan leaders, and then added another five francs for *cahwa*. The effect of that gesture, he noted, was near-magical: "They had just finished telling me a sad and woeful tale of the hardships

The first caravan arrives safely in Biskra.

The young palms get a last drink before their long trip to America.

suffered during that transport; of how they had cut their hands on the offshoots while loading them; of how the camels had turned sideways in the high winds on the desert and had refused to budge; and ending by declaring that they would not go through with another such transport even if I paid them three times the price. But receiving five francs for *cahwa* touched them so that, in parting, they earnestly inquired of me when I would be ready for the next load, and assured me that it was indeed an honor to work for so rich and generous an *Amerikany*, whom Allah surely would give long life and happiness."

And so the laborious and time-consuming task of selecting, inspecting, and preparing the offshoots for overseas transportation had begun. In the following weeks and months, Henry Simon traveled to many other oases, and sent several additional caravans back to Biskra. While engaged in this work, he also tried to identify a single date palm that had proved puzzling to the Department of Agriculture experts back home in California. The tree in question had been imported by the Agriculture Department from Tunisia as an offshoot; they had given this unlabeled palm to a resident of Indio by the name of Johnson. The "mystery palm" developed well and bore fine and delicious fruit, but its characteristics fitted none of the known varieties. Through consultations with some of the most knowledgeable Arab date grove owners, and comparison of the fruit of the unlabeled palm with that of several name varieties, Simon came to the conclusion that the tree probably was a specimen of the highly valued Tafazwin variety; certain slight differences suggested it might have been the offshoot of a very fine Tafazwin seedling.

As the offshoots began to accumulate in Biskra, Simon made arrangements to have them shipped to the coast. For transport to California, the offshoots were packed in large cases, made originally for shipping pianos, and loaded aboard a freighter for New York.

The day arrived when the task was finished, and some 6,000 *djebbar* were on their way to the United States. Ben Ali bid an emotional farewell to his favorite *Amerikany*, and after a brief detour through Europe to visit

Henry Simon in Biskra with 3,000 date palms ready for shipping.

Fourteen thousand newly arrived palm offshoots planted near Mecca.

his parents and his bride-to-be, Henry Simon arrived in New York in time to take charge of his very special cargo, and arrange for shipping it across the continent. Finally, months after they had been checked out in the oases of North Africa, the offshoots arrived at Thermal, in the Coachella Valley, where they were disinfected in a creosol bath. Aware of the disastrous consequences following accidental introduction of foreign plant pests in the past, the government was taking no chances. Thermal was at that time the only station on the west coast where date palms could be unpacked.

As soon as they had been checked out, the offshoots from North Africa and those purchased by Popenoe in the Persian Gulf region— more than 14,000 of them—were planted in tight nursery rows on the land of the West India Gardens. Now began a somewhat anxious waiting period distinguished by unceasing care and labor; the new arrivals not only had to be provided with sufficient water on a strict schedule, but also had to be checked constantly for signs of damage or disease. Soon however, it became evident that all but a few had survived their long trip in good condition. After several months, the young palms began to put down strong roots, and grow their first foliage in their new home.

Henry Simon was in charge of the irrigation system, which he had designed; he and his crew devoted a great deal of time and energy to the project. In the spring of 1914, he brought his young bride to Mecca and to the house he had built for her on his land not far from the plantation. She later recalled the many times her husband got up in the middle of the night and walked across the stretch of desert between the house and the well to check the controls and make sure that the water was flowing properly. The palms flourished, and were subsequently sold to growers all over the valley.

Beginning with that signal year 1913, and through the subsequent ten years—interrupted only by the latter years of the First World War—several companies of date growers that had formed in the Coachella Valley sent their own representatives to the Middle East and North Africa to purchase large numbers of offshoots of different

The irrigation system for the imported palms.

varieties. Many of these palms were brought back by the experienced Bernard Johnson, of early date-growing fame, from journeys similar to the one described by Henry Simon in his travel diary.

The year 1922 signaled the end of the large commercial importations; the only offshoots brought in after that year were small experimental lots of new varieties purchased by representatives of the U.S. Agriculture Department. In the meantime, many of the original companies, including the West India Gardens, were either dissolved or sold. Some of the groves later fell into neglect and were removed, despite the

Henry Simon checks the new growth of a young palm for diseases.

fact that the palms by then were old enough to yield a good harvest, and for some years it seemed that the date enterprise, which was started with such high hopes, was doomed to fail altogether. However, some of the people who purchased the date groves were determined folk, who put in the hard work and persistence needed to overcome all difficulties. Eventually, with thousands of acres planted with palms, and tens of thousands of trees in bearing, the Coachella Valley was transformed into the country's center of date production.

Date grove on the edge of the desert in the Coachella Valley.

6. The Coachella Valley

Home of America's Date Groves

ONE OF THE MOST distinctive features of that portion of California's southeastern desert region known as the Coachella Valley are the huge date groves with their rows of tall, stately trees, whose feathery tops form a delicate lace pattern against the sky and the bare hills that border the valley on both sides. The reasons for selecting this particular American desert area as the New World home of crop-yielding date groves can be found in the peculiar climatic and soil conditions that exist in the Coachella Valley, all of which closely resemble those found in the areas of the North African Sahara where some of the most viable varieties of dates are grown. These environmental conditions, although prerequisite, would not have been sufficient, however, if the valley had not been similar to the North African regions in yet another and all-important feature: the presence of underground water resources, which could be reached by digging or drilling wells in the desert. Such "underground lakes" had created the oases in the Sahara Desert, and were also tapped in the Coachella Valley to supply the huge amounts of irrigation water needed for large commercial date plantations. These efforts signaled a new era for the valley, which until the turn of the century had only a scattering of hardy ranchers, and was for the most part inhabited by the desert flora and fauna that predominated there as well as in the adjoining "high desert" areas of the Mojave Desert. The very special environment found in the low-lying salt regions of the Coachella and Imperial valleys, which together form the so-called Colorado Desert, resulted from a unique combination of the area's ancient geological past and the centuries of continuous action by the Colorado River waters.

Flanked by Mounts San Gorgonio and San Bernardino, whose peaks—each over 10,000 feet high and snow-capped through much of the year—tower above the adjacent mountain ranges, the San Gorgonio Pass forms the entrance to the Coachella Valley that stretches eighty miles beyond to the southeast. From the pass, which reaches an elevation of almost 2,500 feet, the road descends abruptly; at Palm Springs, only twenty-five miles further down, the elevation is a mere 500 feet, and another twenty-five miles along, close to zero. Still the road descends as it passes through small desert towns. Finally, near the large saltwater lake known as the Salton Sea, we find communities that are located 150 feet below sea level.

The northern shores of the Salton Sea mark the boundaries of the Coachella and Imperial valleys; the latter stretches all the way down into Mexico and extends into the Mexicali Valley, which finally ends at the dividing ridge of the Colorado River delta in Baja California. The entire region comprising this chain is known to geologists as the Colorado Desert, or the Salton Basin, as the central part is often called. There are few other areas in the Southwest that can equal, and none that can surpass, this region in interesting features both in regard to its geological past, and in the story of man's conquest of these treacherous yet beautiful desert valleys. Although much less known than the familiar Death Valley, which extends northward from the Mojave Desert, and whose below-sea level location, intense heat, and reputation as a waterless death trap for pioneers and goldseekers of old are interna-

A sand storm moving into the valley.

tionally famous, the Colorado Desert was no less feared by those who knew it.

Spanish conquistador Juan Bautista de Anza, the first white man known to have crossed the Colorado Desert, finally reached San Diego in 1774 with his decimated troops after a disastrous journey. He must have thought back with a shudder to the burning, desolate wasteland they had been forced to traverse between the Colorado River and the mountains to the west, for that desert stretch of a mere hundred miles or so had claimed a disproportionate number of his men. Legend has it that he called the desert *"la jornada de la muerte"*—"the journey of death." Although that description seems to have been favored by the Spanish for any difficult terrain they encountered during their conquests, many who came after de Anza, and who also had to leave many dead companions behind in the hot sands of the Salton Basin, would have agreed wholeheartedly with his words.

The Spaniards had been searching for a land route from northern Mexico to California when they started on their deadly journey. Almost seventy years later, the American general Stephen W. Kearny, who had to cross it in 1847, confirmed de Anza's account on all points. Kearny, whose bloodless conquest during the Mexican War secured for the United States the areas today comprising much of the Southwest, was used to travel in arid regions, but that part of the Colorado Desert impressed him as one of the worst he had ever encountered. Later, during the mad gold rush years of 1849 and 1850, when the gold fever reached its pitch, and vast numbers of people poured into California, many a seeker after gold who took the southern route into the Land of Promise found his last resting place in the desert instead.

The first white settlers came in only after the stagecoach line that linked Yuma and points east with Los Angeles was replaced by a railroad. Yet today, almost exactly two hundred years after de Anza completed his "journey of death," this desert region has become one of the richest, most fertile agricultural areas in the West, yielding to crops that include everything from a variety of vegetables to citrus fruit, grapes, and, of course, the dates that have become the Coachella Valley's most distinctive and famous fruit. All these products are raised in the alkaline soil of what in the ancient geological past was an ocean floor, and even as recently as five hundred years ago the bottom of a large lake.

Going back millions of years in geological history, we find that the entire area and the connecting valleys to the south were part of what is now the Gulf of California, which then extended far northward to form a huge inland sea, probably reaching all the way up to the San Joaquin and Sacramento valleys. There followed tremendous upheavals, which created the mountain ranges that today flank the valleys on both sides, and up from the bottom of the sea came whole areas covered by oyster beds and inhabited by a variety of other marine creatures. This accounts for the Coachella Valley's name, which stems from a misreading, by some Washington official, of the Spanish word *conchella*, meaning "little shell," which in turn refers to the many fossil shells that even today may be found in the desert sands and rocks of that region.

Over the years, the central portions between the mountain ranges settled and formed the valleys, and the remaining salt water evaporated, leaving large salt deposits especially in the lowest-lying areas. Then, very gradually, the Colorado River began to deposit huge quantities of silt eroded from the 240,000 miles of its drainage area, and the thousands of miles of canyons the river was beginning to carve out of the mountains along its course to the north. Here, in the Colorado Desert, much of the silt, eroded from the rock formations that are now visible in the magnificent natural architecture of the Grand Canyon and many others, found its final resting place. Over millions of years, the silt accumulated in layers that geologists estimate as being 12,000 feet deep in some places.

As more and more silt was deposited, the course of the river changed, until finally it was diverted into the Imperial and Coachella valleys, filling the low-lying areas of that ancient evaporated ocean bed with water that formed a vast, 150-mile-long lake, whose shoreline can still be seen clearly today in many places along the mountains, especially in the southwestern part of the Coachella Valley. The deposits known as travertine—crystalline calcium carbonate—are visible as a discolored band sharply delineated against the upper portions of the rock along certain stretches of the foothills leading into the Santa Rosa range near the Salton Sea. The line of demarcation is approximately

Water line of ancient Lake Cahuilla shows along the foothills of the Santa Rosa range.

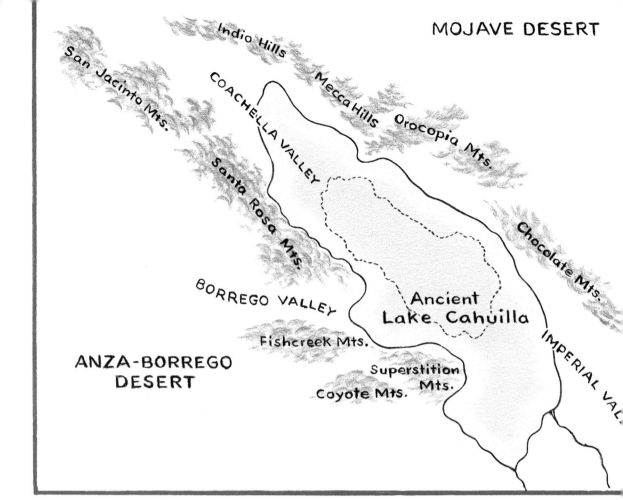

The Salton Basin in the past when much of it was covered by Lake Cahuilla.

thirty feet above sea level; at that point, the ancient lake, which was named Lake Cahuilla, and which in the center had a maximum depth of about 300 feet, would have overflowed across the silt bar and back into the Gulf of California. Bands of Indians living in the canyons and foothills used to fish along the shores of that lake, as shown by the remains of ancient fish traps still visible today on the slopes of the foothills. Although fed by the fresh water of the Colorado River, Lake Cahuilla must have had a considerable degree of salinity, stemming from the old salt accumulations in the soil.

Over the centuries, new silt deposits again changed the flow of the Colorado delta waters, and as the river was diverted, it stopped feeding

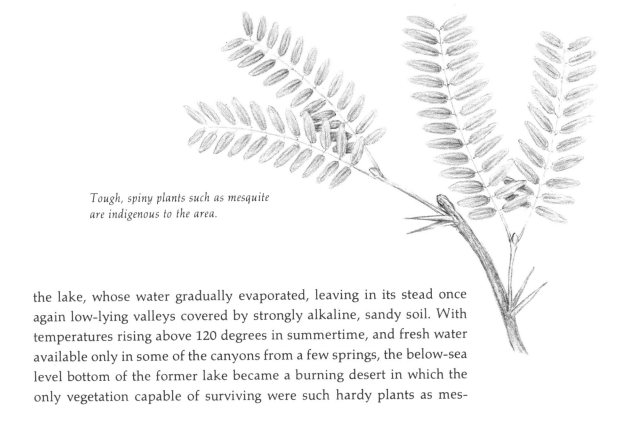

*Tough, spiny plants such as mesquite
are indigenous to the area.*

the lake, whose water gradually evaporated, leaving in its stead once
again low-lying valleys covered by strongly alkaline, sandy soil. With
temperatures rising above 120 degrees in summertime, and fresh water
available only in some of the canyons from a few springs, the below-sea
level bottom of the former lake became a burning desert in which the
only vegetation capable of surviving were such hardy plants as mes-

The Salton Basin as it is today.

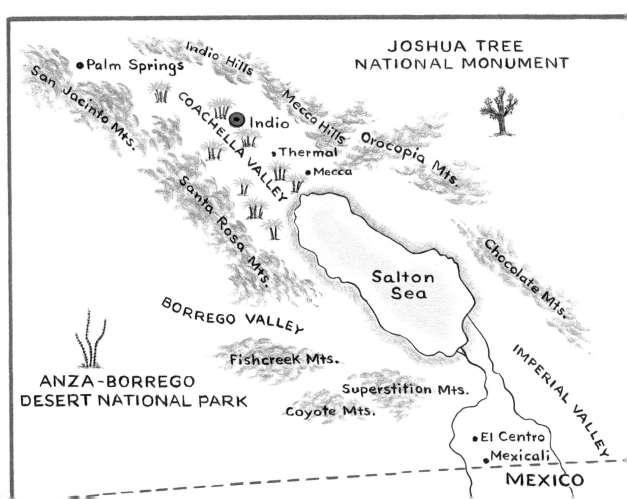

quite, sagebrush, greasewood, and certain kinds of cactus. The handsome Washington fan palms, although also indigenous to that area, remained limited to the canyons where their roots could reach underground water.

We know that for at least five hundred years before the first white men became interested in this region and began to settle in the valleys, there had been no significant overflow into the Salton Basin by the Colorado River, and the valleys belonged to the coyotes, bobcats, jackrabbits, rattlesnakes, and other small desert wildlife. The only human beings to venture into the desert then were occasional bands of Cahuilla Indians bound on hunting trips. Only when the Colorado River rose to unusual heights did some overflow occur, at which times water would inundate the lowest-lying regions beyond the silt bar. We know

Wild-growing Washington palms are found in some of the valley's canyons.

The calm waters of the Salton Sea are ringed by desert and bare hills.

that such overflows occurred during the 1850s, and then again in the 1890s. By the turn of the century, however, all that water had once more evaporated, and the bed of the Salton Sea—at its lowest point some 260 feet below sea level—was so dry that salt was being mined at its northern end.

Only a few years later all that underwent a radical change with the disastrous flood of 1905-07, when the Colorado River broke through the bar, and flooded both valleys. The waters at that time extended as far as Mecca, where they did great damage to cultivated land, including the first experimental date plantations, and threatened settlements even farther north. The flood was finally brought under control by engineers of the Southern Pacific Railroad, to which the task of containing the river had been entrusted. In a monumental effort, entire trainloads of rocks were dumped along an area high and wide enough to form an effective dam, regardless of how much the river rose. What was left of the retreating flood waters accumulated in the lowest-lying basin connecting the two valleys, and so formed the present Salton Sea, a saltwater lake with the approximate salinity of the ocean. About thirty-five miles long and fifteen wide, with a maximum depth of some fifty feet, the Salton Sea has remained more or less constant ever since the flood waters receded almost seventy years ago; if anything, its size has increased somewhat since that time, probably through the runoff water from irrigation.

The coyote

Until the early years of the twentieth century, most of the Coachella Valley retained its desert character and appearance. Tough, spiny plants such as mesquite, cat's claw, and cactus were the predominant vegetation; bobcats and coyotes hunted the lean, long-legged and long-eared desert hares and other, smaller rodents. Some of the reptiles typical of that region were the agile swifts, the desert horned toads, and the peculiar, pale-colored horned rattlesnake, or sidewinder, which made their homes in and among the rocks and scrawny bushes. Most of these animals have special adaptive features that permit them to survive the extremes of a desert environment. Such adaptations may range from the kangaroo rat's unique body chemistry, which converts the dry seeds it

The desert kangaroo rat

The adjoining hills and untouched desert areas have become the last refuge of much of the valley's native wildlife.

eats into water, to the color-change ability of reptiles that helps to protect them against the hot sun.

As man began to cultivate the land, most of the indigenous species became rare or disappeared altogether; a few, however, managed to adapt to the new conditions. Among those that have learned to coexist with man are the wily coyote, the peculiar roadrunner, and the pretty, plumed California quail.

Anyone looking today upon such luxurious resorts as Palm Springs, other flourishing communities, and the thousands of acres of fertile agricultural land in the Coachella Valley, will probably find it difficult to imagine that all of it was still desert as late as Theodore Roosevelt's presidency. Only a few small settlements were found along the railroad; the rest of the non-Indian population consisted of a handful of hardy, homesteading pioneers of the covered-wagon breed. They made a hard-earned living on their ranches with backbreaking work under the blazing sun.

Those today who want to get a picture of what the valley looked like years ago can do so easily by leaving the beaten track and entering the surrounding desert areas still untouched by man. There one finds very little vegetation, and what there is has a brown or grayish-green hue. In

Coachella Valley desert view

the Coachella Valley, the annual rainfall averages only a little over an inch, and that rain may come in sudden great downpours that within a matter of minutes can fill the narrow canyons with six-foot-deep raging torrents capable of moving large boulders, and often sweeping a veritable avalanche of smaller rocks along with them. These downpours cease as suddenly as they begin, the water is absorbed quickly by the sand, and soon afterward, the canyons and the desert lie as barren and dry under the hot sun as before. Only in the spring, after a few brief but

The roadrunner

life-giving rainfalls, does the desert clothe itself in gay, bright colors; now the sand dunes in many places are covered by a thick carpet of wildflowers that seem to have sprung up almost overnight. There are blossoms of purple verbena and yellow desert dandelions mixed with white and green to make a lovely patchwork of colors against the sand and brush. Then the short spring is past, and in the long hot summer that follows, the desert once again lies brown, parched, and apparently lifeless under the hot sun.

Much of the great difference between life in this region now and

A sand lizard

sixty years ago can be expressed in a single word: water. In the early days, the only available water came from wells that had to be dug or drilled in the desert. In order to get a really plentiful supply that could be expected to last, these wells for the most part had to reach deep down into the major water-bearing strata far beneath the desert floor; wells not deep enough to tap such large underground sources often ran dry after a while. Water therefore had to be used as economically as possible; there could be no thought of irrigating vast stretches of cultivated land and also supplying large and sophisticated communities and their expansive modern needs.

All that changed drastically, however, with the realization of an old dream to bring water from the Colorado River to the valleys by way of a canal. This dream finally came true after the completion of the Hoover Dam and the construction of the All-American Canal, which since 1943 supplies water for both domestic and agricultural purposes to the Imperial and Coachella valleys.

By utilizing the water from the canal, the desert land in the center of

The valley's remaining brush-covered areas are still the home of some small wildlife that has adapted to man's proximity.

View from a mature date grove.

the valleys has all but disappeared, and has been replaced by vast stretches of cultivated fields and orchards. The brilliant emerald hues of large, well-irrigated fields of vegetables are alternated by the gold-dotted green of citrus trees, vineyards, and other fruit groves. But most impressive by far, and uniquely typical of the Coachella Valley, are the great forests of date palms with their towering forty- and fifty-foot trunks, which started the valley's agricultural era long before water from the All-American Canal was available. Dates have made the valley famous, and are the one crop that each year attracts thousands of visitors, who come to see the date groves and taste the fruit with such exotic names as Deglet Noor, Barhee, and Medjool. Many of these visitors come because they have heard about the palms from others, and are eager not only to see for themselves and try the dates, but also to learn something about the tree's ancient and romantic history, as well as about the chain of events that brought the date palms to the Coachella Valley many decades ago. Even those with only a casual interest cannot help absorbing at least some information, for groves of fifty-foot trees are difficult to overlook, and reminders of the palm's Old World origin are kept very much alive by all the places that sell dates in the valley.

Proceeding south from Palm Springs on Highway 111—which not without justification is nicknamed the "Date Highway"—the visitor soon sees the first large palm groves appearing on both sides of the road. Especially during the months when the fruit clusters, heavy with ripening dates and usually capped by their protective paper bags, droop from the crowns of the palms, the groves are an impressive sight. In the immediate vicinity of Indio, the self-proclaimed "Date Capital of the United States," one finds not only many large commercial groves, but also date palms planted along roads and streets, in gardens and back-yards. Many of these trees are ornamental plantings only; it is not unusual, however, to see palms bearing carefully protected fruit clusters right on the sidewalks of the town in front of some restaurant, store, or home.

Numerous streets in Indio reflect the preoccupation with the valley's most noted crop; in addition to a Date Palm Avenue, there is a Deglet Noor Street—lined, of course, with palms of that variety—as well as a Biskra, and Oasis, and an Arabia Street.

The date groves that produce practically the entire commercial crop harvested in the United States are concentrated in an area beginning north of Indio and extending south almost to the northern shore of the Salton Sea. Along the highways are found the roadside date shops of the major growers, usually located right beside or even in their groves, where the visitor may sample, buy, and order by mail the popular

*Palms along the roadways
bear their protected fruit.*

varieties, including Deglet Noor, Barhee, and Medjool dates, and try
such unusual treats as date shakes and date ice cream. Here he can also
inspect the showcase gardens some growers maintain especially for the
visitors, and get a close-up view of palms laden with ripening fruit

Deglet Noor Street in the town of Indio.

clusters, or even some young trees that have one or two offshoots as well as several fruit clusters as a demonstration of the palm's two-fold method of reproduction. A walk through a date grove, which some growers also permit, may provide the sight of three or four generations of trees side by side, ranging from newly planted offshoot to full-grown palm.

In the date shops, many people find out for the first time what a really first-class, unprocessed date tastes like, and are almost invariably surprised at the difference in size, texture, flavor, and degree of sweetness of the several available varieties. Through such experiences, not a few casual visitors are turned into aficionados of the ancient staple food of the Arabs, and become regular mail-order customers, for these dates cannot be bought anywhere else in the country. Jealous of their reputation for high-quality fruit, the growers ship by direct mail only in order to guarantee satisfaction.

The Coachella Valley today has an estimated 220,000 date palms under cultivation, which means approximately 4,400 acres, as compared to a total of 260 acres in the rest of the Southwest. These figures do not include many privately owned date palms, whose harvests are used only for home consumption. Many of the old veterans imported from North Africa and the Middle East by Henry Simon, Bernard Johnson, and others still exist and bear fruit; the majority of the valley's palms today are of course the offspring of those old trees.

During harvesting time, which begins in November for some varieties, and ends in February with the late-ripening kinds, one can see in the groves crews of men equipped with long aluminum ladders and other apparatus, including picking belts similar to the devices used by telephone workers, with which the laborer is safely anchored to the trunk while handpicking the clusters. Some varieties warrant harvesting

A fine, mature date grove of one of the valley's major growers.

of the entire cluster at one time; in those instances, the laborer hacks off the cluster after removing the protective cover, and carefully lowers it to a man waiting below at the foot of the palm, who shakes the dates into a large wooden box. They are then sent to the packing plant, where they are gently cleaned, fumigated as a protection against insects that attack stored fruit, graded, and packed. The softer kinds need more care and attention in handling, and are proportionately more expensive. All major growers in the Coachella Valley have their own packing plants, which permits them to control the entire process and guarantee the quality of their fruit.

Although other crops, such as vegetables and citrus fruit, exceed the annual yield of the date crop both in quantity and in cash value, dates are—and probably will remain—not only an important source of local income, but also a tourist attraction that indirectly benefits many other residents as well.

This hopeful prognosis is of relatively recent origin; only a decade or so ago, the date industry appeared to be in serious trouble. Many groves were taken out, or sold to real estate promoters who turned them into motel grounds, residential or commercial building sites; such palms as remained on these lots became purely ornamental trees. Other growers added citrus orchards so that they would not be dependent upon the date crop; there even was some thought of discontinuing large-scale cultivation entirely. In the past few years, however, the growing preference for so-called health, or natural foods, and increased interest in all plant foods also benefited the ancient "bread of the desert." Many grove owners proudly point out that their dates are grown entirely organically, the trees being neither sprayed with insecticides nor treated with artificial fertilizers. Date palms lend themselves well to this kind of natural cultivation because pests attacking the trees are few and rarely become serious threats, and generous irrigation is more important to growth than fertilizers added to the soil. In most cases, turned-under organic matter such as tree cuttings, cover crops, and weeds, plus some manure occasionally is sufficient for maintaining vigorous development and good fruit production. Because no preserva-

Young date garden in the Coachella Valley

tives have to be added in order to keep dates fresh for many months if stored at low temperatures, the claim of the Coachella Valley growers that their dates are an unprocessed "natural food" is essentially correct.

For those tourists who visit the valley in February, there is a special treat in store: the annual Date Festival, which lasts ten days and usually begins in the middle of that month. This custom began in 1921 as a means of paying tribute to the early date pioneers, but was discontinued a few years later. Although revived in 1938, its present format originated in 1948, the year in which the festival for the first time included the outdoors Arabian Nights pageant which since then has become one of its most popular features.

Because the festival has been such a success—it may attract as many

Still-young palm laden with a rich harvest of date clusters.

as a quarter million visitors from all parts of the United States and Canada—the town of Indio built a special, permanent fairground for the annual event, which may best be described as an American county fair with Middle Eastern flavor. The festival starts off with a parade in which Queen Scheherazade and her court of princesses—who are all chosen in annual contests—are the main attraction, but which also features groups attired in fanciful Mideastern garb, as well as colorful Mexican costumes and many handsome Western horses. At the fairgrounds, there is a nightly performance of the Scheherazade pageant, and some unusual attractions, including camel and ostrich races. Aside from the agricultural displays, which of course feature showcase samples of different date varieties, the stone and mineral exhibits are among the best in the

United States, and many stalls offer handcrafted items such as turquoise and silver Indian jewelry, as well as unfinished stones.

Throughout it all, there are constant reminders of the festival's origin. Harem girls, Bedouins, and sheiks can be seen wandering about, for many of the town's residents enter into the spirit of the occasion and dress up in what passes there as Middle Eastern garb. It is not at all unusual to come into a local store—or even a bank—and find the clerks and tellers dressed as emirs and harem beauties.

Readying a grove for the harvest

A worker atop a 40-foot ladder removes a fruit cluster.

Amid the crowds, the excitement, and the pageantry of the Date Festival, there is little chance to reflect on how it all began. But wandering about in the date groves and looking up at the tall trees, one cannot help being impressed by the continuity of life they represent. These palms, propagated through centuries by offshoots cut from the trunks of the parent trees, are quite literally wood from the wood of those that grew thousands of years ago in Mesopotamia, the avowed location of Paradise—which, after all, was a garden.

In his autobiographic reminiscences, Henry Simon closes the part dealing with his North African adventures with a reflection on the plantations he helped to start. "Today there are great forests of date palms in the Coachella Valley," he writes. "Most of the original palms I bought in the oases of the Saraha are still there. No one would recognize in these towering trees, which together with their descendants supply a significant portion of the dates harvested in the United States, the small, dry-looking *djebbar* I marked with my signature so many years ago in the courtyards of North African oases."

Photographic Credits

Index

Pages on which illustrations appear are shown in *italics*